Applying For Patents

Protecting Your Intellectual Property

Calvin Lowe

Straightforward Publishing

D1321683

Straightforward Guides

British Library Cataloguing in Publication data. A Catalogue record of this book is available from the British Library.

ISBN

978-1-84716-654-8

Printed by 4edge www.4edge.co.uk

Applying For Patents

Protecting Your Intellectual Property

Contents

Introduction

This Second edition of Applying for Patents-Protecting Your Intellectual Property, is a concise introduction to all aspects of patents, trade marks, regsitered and unregistered designs and copyright.

The book deals comprehensively and clearly with a complex, elusive and rapidly changing area, of importance to those engaged in the commercial world, or to teachers of the subject.

The law of intellectual property impinges upon the lives of many people, whether traders, artists, musicians or designers. Whatever we do, whatever we create, we need to understand what protection the law affords our endeavours. We also need to understand the practical steps involved in applying for protection.

The following chapters contain a discussion about the nature of intellectual property generally and then covers the application process (where applicable) in depth, enabling the reader to progress from a point of lack of understanding of intellectual property protection to successfully navigating the path through to obtaining some form of protection.

The processes are quite often long and can be costly, but in the increasing competitive world of business, protecting ones own creation is essential.

Most of what is required can be done on an individual basis without the intervention of costly patent attorneys. The website of the Intellectual property Office www.ipo.gov.uk provides an invaluable source of information.

Ch.1

Patents Generally

Why is it you are choosing to apply for a patent? Many people waste a lot of time and money applying to protect something that, in the end, can't be protected for one of a number of reasons. The information below will help you to consider the reasons that you are applying for patent protection and chapter 2 provides a step-by-step guide to making the application.

What does a patent protect?

A patent protects new inventions and covers how things work, what they do, how they do it, what they are made of and how they are made. If a patent application is granted, it gives the owner the ability to take legal action under civil law to try to stop others from making, using, importing or selling the invention without permission. www.gov.uk/government/organisations/intellectual-property-office provides guidance on resolving patent disputes. The Intellectual Property Office has very detailed guidance on applying for patents, trademarks, registered designs, copyright and all other forms of protection.

The Intellectual Property Office (IPO)

The IPO is the official government body responsible for Intellectual Property (IP) rights in the United Kingdom. These rights include:

- Patents
- Designs
- Trade marks
- CopyrightOther forms of protection

The IPO is an Executive Agency of the Department for Business Innovation and Skills. They promote innovation by providing a clear, accessible and widely understood IP system, which enables the economy and society to benefit from knowledge and ideas. As an Executive Agency of the BIS they are directed and controlled by corporate governance rules to ensure that they supply public services of the highest quality, share good ideas, control costs and above all deliver what they are supposed to. To ensure that they act within these rules Ministers appoint independent members to sit on a Steering Board. This Steering Board has no executive functions; its role is to advise Ministers, through the Director General, on the strategies that the IPO must adopt.

Patent Attorneys-What is a patent attorney?

A patent attorney is a member of a profession of around 2000 members who have a particular expertise in the field of intellectual property. Intellectual property encompasses patents, industrial designs, design rights and related copyright areas, from computer programs to the shape of teapots, and trade marks. Patent attorneys work either in patent departments of large industrial organisations, in private firms of patent attorneys, or in government departments, and their work deals with obtaining and enforcing intellectual property rights.

Qualifications
To start with, all patent attorneys need a scientific or technical background. Nowadays, this usually means a science or engineering degree from a university or similar institution that confers eligibility for both UK and European qualifying examinations. A scientific training, however, is not enough. The would-be patent attorney must have the ability to acquire, and enjoy exercising, legal skills of

drafting, analysis and logical thought, and, particularly the skill to use the English language aptly and accurately, in written work. In addition, because patents are increasingly international, a knowledge of at least French and German, although not essential, is highly desirable. Patent attorneys act at an exciting interface between disciplines of law, language and science.

Recent legislation has granted patent attorneys the same rights as solicitors and barristers to conduct litigation (i.e. to control the conduct of the cases) and to act as advocates in the Patents County Court. This Court was set up to hear patent and other intellectual property cases without the complexity and cost of High Court proceedings. Many patent agents become heavily involved in litigation generally in the Courts of the UK and in other countries. Also, some patent attorneys acquire an additional qualification entitling them to conduct litigation in the High Court.

The history of patents

The origins of patents for invention are obscure and no one country can claim to have been the first in the field with a patent system. However, Britain does have the longest continuous patent tradition in the world. Its origins came from the 15th century, when the Crown started making specific grants of privilege to manufacturers and traders.

Open letters marked with the King's Great Seal called Letters Patent, signified such grants. Henry VI granted the earliest known English patent for invention to Flemish-born John of Utynam in 1449. The patent gave John a 20-year monopoly for a method of making stained glass, required for the windows of Eton College that had not been previously known in England.

Tudors and Stuarts

In the time of the Tudors, it became common practice for the Crown to grant monopolies for trades and manufacturers, including

9

patents for invention. From 1561 to 1590, Elizabeth I granted about 50 patents whereby the recipients were enabled to exercise monopolies in the manufacture and sale of commodities such as soap, saltpetre, alum, leather, salt, glass, knives, sailcloth, sulphur, starch, iron and paper.

Under Elizabeth I and her successor James I, the granting of monopolies for particular commodities became increasingly subject to abuse. It was common for grants to be made for inventions and trades that were not new. In some instances, grants were made to royal favourites for the purpose of replenishing royal coffers.

In 1610, James I was forced by mounting judicial criticism and public outcry to revoke all previous patents and declare in his "Book of Bounty" that 'monopolies are things contrary to our laws' and "we expressly command that no suitor presume to move us". He stated an exception to this ban for "projects of new invention so they be not contrary to the law, nor mischievous to the State".

The doctrine of the public interest was introduced into the patent system and the words were incorporated into the Statute of Monopolies of 1624. Section 6 of the Statute rendered illegal all monopolies except those "for the term of 14 years or under hereafter to be made of the sole working or making of any manner of new manufactures within this Realm to the true and first inventor".

The 18th century

In the 200 years after the Statute of Monopolies, the patent system developed through the work of lawyers and judges in the courts without government regulation.

In the reign of Queen Anne, the law officers of the Crown established as a condition of grant that "the patentee must by an instrument in writing describe and ascertain the nature of the invention and the manner in which it is to be performed".

James Puckle's 1718 patent for a machine gun was one of the 1st to be required to provide a "specification". The famous patent of

Arkwright for spinning machines became void for the lack of an adequate specification in 1785, after it had been in existence for 10 years.

Extensive litigation on Watt's 1796 patent for steam engines set out the important principle that valid patents could be granted for improvements in a known machine. It also established that a patent was possible for an idea or principle, even though the specification might be limited to bare statements of such improvements or principles, provided they come into effect, or were "clothed in practical application".

The 19th century

Britain's patent system served the country well during the dramatic technological changes of the industrial revolution. However, by the mid-19th century it had become extremely inefficient. The Great Exhibition of 1851 accelerated demands for patent reform.

Up to that time, any prospective patentee had to present a petition to no less than seven offices and at each stage to pay certain fees. Charles Dickens described the procedure in exaggerated form, somewhat derisively, in his spoof, "A Poor Man's Tale of a Patent", published in the 19th-century popular journal "Household Words"; Dickens' inventor visits 34 offices (including some abolished years before).

The Patent Office came about to meet public concerns over this state of affairs, and was established by Patent Law Amendment Act of 1852. This completely overhauled the British patent system and laid down a simplified procedure for obtaining patents of invention. Legal fees were reduced and the publication of a single United Kingdom patent replaced the issuing of separate patents for each nation of the Union.

A subsequent Act in 1883 brought into being the office of Comptroller General of Patents and a staff of patent examiners to carry out a limited form of examination; mainly to ensure that the specification described the invention properly.

An important milestone in the development of the British patent system was the Act of 1902, which introduced a limited investigation into the novelty of the invention before granting a patent. This required patent examiners to perform a search through United Kingdom specifications published within 50 years of the date of the application. Even with this restricted search, a vast amount of preparatory work was involved and an additional 190 examiners assisted the existing staff of 70 examiners.

By 1905, to enable searching, patent specifications from 1855 to 1900 had been abridged and classified in 1,022 volumes arranged in 146 classes according to subject. By 1907, the abridgement volumes extended back to the first patent to have a number:

- Patent No. 1 of 1617 granted to Rathburn & Burges for "Engraving and Printing Maps, Plans &c".

The legislation in force at present is the Patents Act 1977, as amended by the Intellectual Property Act 2014. The 1977 Act was the most radical piece of patents legislation for nearly 100 years. The Act sets out to ensure that the patent system is well suited to the needs of modern industry, sufficiently flexible to accommodate future changes in technology and adapted to operate in an international context. The changes introduced in the IPA 2014, have been made broadly to:

- make it easier for patent owners to provide public notice of their rights
- provide for the agreement establishing a Unified Patent Court to be brought into effect in the UK

- allow the Intellectual Property Office (IPO) to share information on unpublished patent applications with international partners to speed up patent processing and
- expand the circumstances in which the IPO may issue an opinion in respect of patents

Main changes to patents

- Easier notification: Instead of including a patent number on a product, a patent owner can now put others on notice of his rights by including a link to a website which provides details of the patent
- Increased powers of IPO: as broader power to give opinions and the Comptroller General of Patents, Designs and Trademarks has greater power to revoke a patent on his own initiative
- Single European patent: The act enables the inter-governmental agreement to provide for a Unified Patent Court within participating European countries, to be brought into effect in the UK. This will mean that it will be possible to apply to the European Patent Office for a single patent which has effect across all participating countries and for some issues to be litigated in a new Unified Patent Court. This could save businesses a considerable amount of money
- Permits information sharing: The act allows the IPO to share information on unpublished patent applications with international partners, in certain circumstances, to speed up patent processing
- Minor amendments: There are a number of other minor amendments made to the Patents Act 1977 including extending the period during which a third party can challenge ownership of a granted patent

Ch. 2

Obtaining Patent Protection

Vey importantly, to obtain patent protection, your invention must:

- be *new (novelty)*
- have an *inventive step* that is not obvious to someone with knowledge and experience in the subject
- be capable of being made or used in some kind of industry

However, your invention must not be

- a scientific or mathematical discovery, theory or method
- a literary, dramatic, musical or artistic work
- a way of performing a mental act, playing a game or doing business
- the presentation of information, or some computer programs
- an animal or plant variety
- a method of medical treatment or diagnosis
- against public policy or morality.

If your invention meets the above requirements, then you will be eligible to apply for a patent.

Novelty and the Inventive Step

As we have seen, when applying for a patent, an invention must satisfy three main requirements. Firstly, it must be inherently patentable (i.e. not part of the unpatentable subject matter defined by the UK Courts). If it fulfils this first requirement, an invention must then be both new (i.e. it must not have been previously disclosed to the public) and involve an inventive step – in other words, it must not be an obvious modification of what already exists. Typically, the hardest of these requirements to overcome is to show that an invention is not obvious and, unsurprisingly, shows a degree of invention.

Section 3 of the UK Patents Act 1977 states "An invention shall be taken to involve an inventive step if it is not obvious to a person skilled in the art, having regard to any matter which forms part of the state of the art"

The majority of patents are improvements on technical solutions that already exist. These improvements may be small or big – obvious or non-obvious. The obviousness of the improvement is judged according to the knowledge of a notional "person skilled in the art". For example, for an invention concerning a new chemical, the "person skilled in the art" will be a chemist; a mechanical engineer is the skilled person for a mechanical invention. The state of the art is everything that has been made available to the public, through either written or oral description, before the date of filing of the patent application.

The pitfalls of not patenting an invention
The pitfalls of not patenting your invention are immediately obvious. If you choose not to patent your invention, anyone can use, make or sell your invention and you cannot try to stop them. You can attempt to keep your invention secret, but this may not be possible for a product where the technology is on display.

The benefits of applying for protection
Most importantly, a patent gives you the ability to take legal action to try to stop others from copying, manufacturing, selling, and importing your invention without your permission. The existence of your patent may be enough on its own to stop others from trying to

exploit your invention. If it does not, the patent gives you the right to take a legal action under civil law to try to stop them exploiting your invention.

A patent will also allow you to:

- sell the invention and all the intellectual property (IP) rights
- license the invention to someone else but retain all the IP rights
- discuss the invention with others in order to set up a business based around the invention.

The public also benefit from your patent because the Intellectual Property Office publishes it after 18 months. Others can then gain advance knowledge of technological developments which they will eventually be able to use freely once the patent ceases.

Before you apply

There are a number of important considerations which need to be taken on board before applying for a patent, not least the cost of the application, which are paid in several stages..

Application fee

The application fee GBP £30 (GBP £20 if filed using the e-filing or web-filing service) relates to the preliminary examination of an application and is not a filing fee. Once the IPO have done the preliminary examination of your application, they won't refund the application fee for any reason.

The initial filing fee is minimal. There are more important questions to consider before making your application.

Ask yourself the following:

Have you considered other forms of protection?

There may be other forms of protection that are more suitable for your invention, instead of or in addition to patent protection.

Does your product meet the requirements for protection?

To be eligible for patent protection, your invention must be new, inventive and must not be of an excluded type.

Is your invention new?

The IPO will not grant a patent if your invention is public and isn't new. You should try to find it elsewhere before applying for a patent.

Are you the legal owner of your invention?

A patent applicant may be an individual or a corporate body, and persons (whether individuals or corporate bodies) can make joint applications.

Have you received enough advice?

You can get confidential advice regarding your proposed application from a number of different sources, such as patent attorneys, solicitors, or staff at the Intellectual Property Office.

Have you considered patent protection abroad?

A United Kingdom (UK) patent is a territorial right that only gives protection in the UK. Check out the alternatives if you are thinking of protection abroad.

Permission to file a patent application abroad may be required in some cases.

Search existing patents

You can search Espacenet (worldwide.espacenet.com) to see if your idea is new. Note that this service is not designed as a comprehensive patents search facility. Therefore, it should not be used to determine conclusively whether your idea can be patented.

You can get confidential advice regarding your proposed application from a number of different sources, such as patent attorneys, solicitors, or the IPO staff.

Are you the owner or inventor?

If you are the creator of an invention, you are usually regarded as the owner and entitled to apply for patent protection. However, if you created the invention as an employee, in the course of your normal duties, the invention belongs to your employer.

A patent applicant may be an individual or a corporate body, and persons (whether individual or corporate) can apply for a patent

An application for a patent should include a full description of your invention (including any drawings), a set of claims defining your invention, a short abstract summarising the technical features of your invention and a filled in form 1 (see later).

Claims

A claim is a definition in words of the invention that you want to protect.

Writing your claims

Your main claim should list all of the main technical features of your invention including those that distinguish it from what already exists. Subsidiary or preferred features, which are not crucial to your invention, should be set out in dependent claims that can refer to one or more of the previous claims. You must write each claim as a single sentence.

Do not include any statements relating to commercial or other advantages or other non-technical aspects of your invention in your claims.

You should file the claims and description at the same time.

Although you may choose to file your claims later, it is recommended that you file them with your application because: later filed claims will not be allowed to encompass any matter which is not disclosed in your description; if you do not file claims in time your application will be terminated; and without claims the IPO are unable to search your application. Later filed claims may be filed online.

Description

Your description puts into words the features of your invention and must contain enough information for others to carry out your invention. You must explain your invention fully in your description when you first give it to the IPO because you will not be allowed to add information later. For this reason it is important to explain the full scope of your invention.

How to produce your description

Your description should be typed or printed on one side of separate sheets of plain white A4 paper with 2cm margins on all sides. It should have a short title, which shows the general subject of your invention. It may typically begin with the background of the invention and if appropriate explain the problem it solves, and what the invention does. It should then set out the key features of your invention and any important but not so key features. It should then include a brief introduction to your drawings. The rest of your description should show in more detail with reference to your drawings one or more particular examples of your invention.

When is your description needed?
The IPO will need your description with a Patents Form 1 to give you a filing date

Drawings
Your application may include drawings showing the technical features and construction of your invention to help understand the description and assist in understanding your invention. They may illustrate different views of your invention as seen from different angles, or if appropriate, cross-sectional views. You should include drawings of any important features hidden in use. The drawings help people to understand your invention and should illustrate the range of ways that it can be put into practice.

How to produce your drawings
Your drawings, which must use black, well-defined lines, should be on one side only of separate sheets of plain white A4 paper with margins of at least 2cm at the top and left-hand side, 1.5cm at the right and 1.0cm at the bottom.

You should indicate specific features in the drawings using reference numbers and or letters connected to the features with clear, continuous lines. You should use these references in the description to refer to those features. You should use the same reference number or letter to refer to the same feature in all the figures that show it.

Each drawing sheet may carry one or more drawings. You should label the drawings sequentially as "Figure 1", "Figure 2" and so on. Number the sheets in order if you have more than one sheet of drawings, for example, with three sheets use "1/3, 2/3, 3/3".

You should send your drawings at the same time as you give the IPO your description

Fill in form 1

Form 1 is the main form that you send in with your claim, description and drawings. Fill in form 1 in capital letters using black ink, or type them.

You can also send your request for search and fee sheet with your application. If you decide not to do this now you will be informed of the date by which you must send them to ensure your application continues. This will be no later than 12 months from your filing date.

If you are not the inventor, there is more than 1 inventor or you are applying on behalf of a company, you must also send a statement of inventorship.

Sending in the form

You should send the filled in application form, description, drawings, claims, abstract and statement of inventorship form, if necessary, to:

Intellectual Property Office
Concept House
Cardiff Road
Newport
South Wales
NP10 8QQ
United Kingdom

What happens next?

When the IPO receive your completed application form, description and drawings they will send you a receipt, within 3 days, confirming the date they received your application and an application number.

Your application details including your name and address will appear on their records. They also include them in their searchable patents journal when they publish your application. Both are available to the public on their website and can be permanently searched using most standard search engines. If you do not want your home address published, give them a different permanent address where you can be contacted, such as a business address or a suitable PO Box address.

All correspondence from yourself, the Office and any third parties will be open to public inspection, including on their website, once your application is published. Indviduals or corporate bodies can make joint applications.

Confidentiality

It is important that you do not make your invention public before you apply to patent it, because this may mean that you cannot patent it, or it may make your patent invalid.

However, that does not mean that you must never discuss your invention with anyone else. For example, you can discuss it with qualified (registered) lawyers, solicitors and patent attorneys because anything you say to or show them is legally privileged. This means it is in confidence and they will not tell anyone else.

Alternatively, you may need to discuss your invention with someone else before you apply for a patent – such as a patent adviser or consultant, or an inventor-support organisation. If so, a Non-Disclosure Agreement (NDA) can help. NDAs are also known as confidentiality agreements and confidentiality-disclosure agreements (CDA).

No single NDA will work in every situation. This means that you must think carefully about what to include in your NDA. You may want to consult a qualified lawyer or patent attorney if you are thinking about discussing your invention with someone else and are considering using a non-disclosure agreement.

The IPO's booklet, Non-Disclosure Agreements (NDA) gives information and guidance about what you need to consider when disclosing an invention, including example NDA templates.

Statement of inventorship

When you apply for a patent you need to give the IPO more information about who invented it if:

- you are not the inventor
- you are not the only inventor
- you are applying on behalf of a company

If you are not the inventor or are applying on behalf of a company, you must also tell them how you have the right to apply for the patent.

How much does it cost?

Most people are put off the idea of applying for a patent because of the cost, or potential cost. If you use a patent attorney then for sure you will pay a lot of money. However, it is relatively inexpensive to apply yourself. The normal amount charged to process a UK patent application is GBP £230 - £280.

As stated, If you decide to seek professional IP advice (from a Patent Attorney or other representative) you will need to factor in the cost of this as well. If the patent is granted, you must pay a renewal fee to renew it every year after the 5th year for up to 20 years protection. Renewal fees start at £70 for the 5th year and rise to £600 for the 20th year.

Paper filing

- GBP £30 (application fee) for a preliminary examination
- GBP £160 for a search
- GBP £80 for a substantive examination

Electronic filing/web-filing service

- GBP £20 (application fee) for a preliminary examination
- GBP £130 for a search
- GBP £80 for a substantive examination

How to pay

If you apply on-line, you can pay by credit or debit card, or by deduction from your deposit account (if you have one-see below) with the IPO. If you apply by post, you can also pay by cheque or bank transfer, but you must also fill in and send a Form FS2 fee sheet with your application. See the IPO website for further details of paying by transfer.

Deposit account

If you regularly do business with the IPO, you can pay by deposit account. This is particularly useful if you have to meet any last-minute payment deadlines.

After you apply

After you apply, the IPO will:

- check your application meets their requirements
- send you a receipt with your application number and filing date, this is the date they receive your application
- tell you what you need to do and when.

You must send claims, abstract, application fee and search fee within 12 months of your filing date, or priority date.

Request search

You must request a search within 12 months of your filing or priority date. The IPO will check your application against published patents and documents to check your invention is new and inventive.

Publishing an application

If your application meets their requirements, they publish it just after 18 months from your filing or priority date. They will also make many documents on the open part of the file for the patent available to the public including by putting them on their website.

Your name and address will appear on the front page of the published application. Upon publication these details will also appear in IPO records and in their online Patents Journal, both of which are available to the public on their website and can be permanently searched using most standard search engines. If you do not want your home address published, you should give a different address where you can be contacted, such as a business address or a PO Box address.

Request substantive examination

You must request substantive examination within 6 months of publication. The IPO will examine your application and tell you if it meets the legal requirements. If it does not they will tell you what you need to do and how long you have to do it. This can continue for up to 4½ years from your filing or priority date.

Accelerated procedure

Various methods of accelerating the examination procedure are available - see the IPO patents fast grant guidance for further details. For example, if your application relates to an invention with an environmental benefit, accelerated processing is available through the Green Channel for patent applications. Acceleration options are also provided by the Patent Prosecution Highway and PCT (UK) Fast Track, each of which allow intellectual property offices to make use of work already conducted at another office.

Granting a patent

If your application meets the strict legal and technical requirements, the IPO will grant your patent, publish it in its final form and send you a certificate. You then need to pay renewal fees each year to keep the patent in force.

A typical patent application takes 3 to 4 years to grant, however the procedure may be accelerated as explained above. There is generally a time limit of 4½ years from the application's earliest date. You must meet requirements within the given time limits, or your application may be terminated. However, you can extend some time limits.

If your invention is pharmaceutical or a plant protection product, you may be able to extend your patent protection with a supplementary protection certificate (SPC).

Disputes

Hearings

Typically there are two types of dispute that you might find yourself in:

- a disagreement with the IPO regarding an objection raised against your patent application or patent, or
- a disagreement with someone else about a patent, for example an act of infringement or dispute about ownership.

If you find yourself in either of these situations, there are several ways in which the IPO can help you resolve your dispute.

Request a hearing

...to resolve a dispute between you and the Office

Sometimes, while the IPO is looking into your patent application or granted patent, they might have to object to certain things about it. You will always be given a chance to overcome these objections, but that may not always be easy. If that happens, you can ask for the matter to be referred to a senior officer at a hearing where you will be given the chance to present your arguments in person.

...to resolve a dispute between you and someone else

You may also find yourself in dispute with someone else over a patent or patent application. If this happens, the IPO might be able to help settle the dispute, but only if you refer the matter to them. If you do, then they will give both sides an opportunity to put their case to a senior officer at a hearing, often known as an "inter partes hearing". Having heard both sides of the argument, the hearing officer will issue a decision that is binding on both parties.

Alternative methods of dispute resolution-Request an opinion

If you are involved in a dispute with someone else about infringement of a patent or the validity of a patent and want to try to resolve this without getting involved in full legal proceedings you might want to consider asking the IPO for an opinion.

Mediation

Mediation is a form of alternative dispute resolution ("ADR"). It allows opposing parties to discuss the dispute with a mediator. The mediator will facilitate an exploration of the issues behind the dispute as well as any possible solutions. There are many benefits to mediation, including being able to resolve disputes spanning several countries. Mediation also offers the opportunity to explore other non-patent related disputes at the same time. To provide parties with an opportunity to mediate the IPO have set up a mediation service, details on their website.

Patent protection abroad

United Kingdom (UK) patents only give you protection in the UK, so you should consider protection abroad as well. Permission to file a patent application abroad may be required in some cases. Before considering protection abroad, ask yourself the following questions:

Do you want to sell your invention abroad?
You may not want to do this now, but you need to think ahead and decide if this is a possibility in the future.

Do you want to license your patent abroad?
This could prevent unlicensed copying or use of your invention.

If you answered no to both questions, you probably do not need to apply abroad. However, please remember this allows anyone to legally make, sell or use your invention abroad.

If you answered yes to either question, you should consider which option for protection abroad works best for you.

Option 1: Extend your patent

Some countries may allow you to extend your UK patent, and accept it as protected in that country after completing certain local formalities. More information about extending your UK rights abroad can be found in the Professional Section of the IPO website.

Option 2: Apply to individual national patent offices

Apply to individual national patent offices if you want protection in individual countries. You can use a certified copy to prove details about your patent when applying.

Apply under the Patent Co-operation Treaty (PCT) to countries worldwide

Apply under the European Patent Convention (EPC) to countries in Europe

Option 3: Use a combination of routes

You can use a combination of routes to apply for protection abroad. For example, you can apply in a single country, including the UK, and apply later elsewhere using the first application to claim priority.

Should you proceed with your patent application or withdraw it?

Once your patent application is published, the contents of your application and all correspondence on file will be in the public domain. It will not be possible to reverse this. Therefore you should consider carefully whether you have reasons for withdrawing the application before publication takes place. Withdrawal is the only way to prevent publication of your application, and the earlier you can notify the IPO of this, the better. You should consider the following factors before deciding whether you want your application to be published or whether withdrawal might be a better option for you.

Why publish?
- Your application has to be published if you want a patent to be granted.
- Publication can prevent other people from patenting a similar invention, even if you choose not to proceed with your own patent application.

Why withdraw prior to publication?
- If there is information on the file for your application which you do not wish to be made public:
- All correspondence from yourself, the Office and any third parties will be open to public inspection, including on the IPO website, once your application is published. If there is

information on the file which you do not want to be published, then you should withdraw your application before publication.

This information could include sensitive information such as letters containing personal information about yourself or others, further details about your invention e.g. amended pages or claims filed after the filing date and containing additional matter, or letters giving further details explaining why your invention is different from the prior art and disclosing details.Or other information which you do not wish other people to see.

If you wish to file abroad at a later date: You will not be able to get a patent granted in another country if the invention has already been made public. If you wish to file a patent application in another country you should therefore either do so before publication of your current application or withdraw your current application before publication.

If you wish to keep your invention secret: You may not want other people to know what you are working on. You may decide that you want to keep the whole invention secret and rely on "trade secrets" rather than obtaining a patent.

If you have lost interest: You may decide not to pursue the patent application, through having lost interest, being unable to manufacture the invention or being unable to get financial backing.

Important changes to patent rules from October 2016

It is important to note that from october 2016, and April 2017, certain fundmental changes will be introduced in the patent process which should be noted..

Notification of intention to grant

On 1 October 2016, the Intellectual Property Office (IPO) will start to issue a new notification of intention to grant. This letter will inform a patent applicant that their application meets all the requirements for grant of a patent. It will also state that the patent application will be sent for grant on or after a certain date. No further action will be required from the applicant in relation to the patent application in question. However, the letter will provide applicants with a clear picture of the time left to take any other action they wish before grant. In particular, there will be a clear timeframe within which to file any divisional patent application.

In the majority of cases the letter will give one month's advance notice of grant. Two months' notice will be provided if the letter is issued at the first examination, as is currently the case. (This two months' notice is required by the Patents Rules. It ensures the applicant has an opportunity to make voluntary amendments to their patent application or file a divisional application.) The introduction of these notifications will end the practice of "foreshadowing" divisional applications. (That is to say, the practice of applicants raising the possibility of filing a divisional application, and asking for time to do so before a patent is granted.) Instead, from 1 October, applicants will be guaranteed a period to file such applications. They will know that the main application has met all the requirements for grant, but will have time before its actual grant.

"Intention to grant" notifications will be issued from 1 October 2016 onwards. The introduction of this change in practice means that very few patents will be granted during October 2016. Patents granted on Tuesday 4 October 2016 will be the last patents to be granted with no advance notification of grant. Notifications of intention to grant will be issued on applications which have met all the requirements for grant of a patent by the compliance date. The notifications will form a report under section 18(4) of the Patents Act. They will be publicly available on the IPO's online patent information site.

In order to introduce this change, rule 19 of the Patents Rules will be amended on 1 October 2016. It will allow divisional applications to be filed after a notification of intention to grant has been issued. The requirement that divisional applications must be filed at least three months before the compliance date will be retained. It will also continue to be the case that divisional applications must be filed before grant of the earlier application.

Simplifying the time period for requesting reinstatement

Where a patent application is terminated because the applicant unintentionally failed to meet a deadline, it is possible to request reinstatement of the application. However the current deadline for requesting reinstatement is complicated to calculate and causes confusion. The deadline will change on 1 October 2016. Applicants will simply have 12 months from termination of their patent application to request reinstatement. The new rule will be applicable to all terminated patent applications from 1 October 2016 onwards. This means that, from 1 October 2016, reinstatement may be requested in relation to any patent applications that have previously been terminated as long as no more than 12 months has passed since the date of termination. It will remain possible for the IPO to impose third-party terms when reinstating a patent application. These allow anyone who has proceeded to make or use the invention, on the basis of the application being terminated, to continue to do so.

Allowing extensions to the period for providing an address for service

All patent applicants must provide an address for service in the European Economic Area or Channel Islands. This enables the IPO or other parties to contact them about their patent application or patent. Currently the period for providing this address is two months and the period is not extendable. From 1 October 2016 applicants will be able to request a two-month extension to this period. This can be done by filing Patents Form 52 accompanied by the relevant fee (currently £135). The request for an extension must be made no more than two months after the period has expired.

Relaxing the requirements for formal drawings

On 1 October 2016 the legislative requirements relating to technical drawings included in patent applications will be relaxed. They will allow shading in drawings, providing the shading does not obscure other parts of the drawing. It will also be possible to include black and white photographs, providing they are clear and capable of reproduction. This will bring the legislation more clearly into line with current IPO practice, providing clarity and legal certainty for patent applicants. The IPO will continue to reject drawings and photographs if they are not suitable for publication.

Amending international patent applications upon entry to the UK national phase

International patent applications are initially processed by the World Intellectual Property Organisation (WIPO) under the Patent Cooperation Treaty (PCT). The applicant can then choose to continue the patent application process in individual jurisdictions. This part of the process is known as the "national phase". In the UK national phase, applicants can amend their patent application voluntarily in response to reports made by the international authorities. This deals efficiently with issues raised by the international authorities, speeding up the application process as a result.

The change on 1 October 2016 will clarify when an applicant can amend their international patent application which enters the UK national phase. The first scenario is where an international search report (ISR) has been issued by the time the application enters the UK national phase. Here, the applicant can amend voluntarily from the time of UK national phase entry up until the first UK examination report is issued. The second scenario is where no ISR has been issued by the time the application enters the UK national phase. Here, the applicant can amend voluntarily from the date of issue of the UK search report or the ISR (whichever is first). Again, the cut-off is the date of issue of the first UK examination report. This rule change will bring the legislation more clearly into line with current IPO practice. It will also provide clarity and legal certainty for patent applicants. It will remain possible to amend all patent applications voluntarily in response to the first examination report, and at other times with the IPO's permission.

Clarifying requirements concerning changes of names and addresses

This change concerns name or address details registered with the IPO in relation to a patent application or a patent. The law is currently unclear about what requirements apply when a customer wishes to update this information. Such an update will be necessary when the information was previously correct but there is then a change in situation. An example may be a house move or a change of name. The legislation is being clarified to make clear that the procedure relating to corrections of names and addresses (used when there was an error in the information recorded) also applies where there has been a change in circumstances. Patents Form 20 will continue to be the form which is used to correct or update a name or address. The form will be revised on 1 October 2016 to include a new tick box. This will allow the user to specify whether they wish to correct an error or to update information held by the IPO.

Advertising amendments made during proceedings

This change concerns the situation where the validity of a granted patent is challenged during infringement or revocation proceedings. The patent proprietor may request to amend the patent in order to avoid revocation. The legislation is clarified on 1 October 2016 to reflect that, under the Patents Act, the IPO has discretion over whether to advertise such amendments. In practice the only amendments not advertised would be those which are so insignificant that no-one could be expected to want to oppose them. Amendments to a patent which are proposed to avoid revocation are very unlikely to be insignificant. It follows that these amendments will be advertised. Guidance such as the Manual of Patent Practice will set out the factors the IPO will consider before deciding whether to advertise or not.

Removing certain requirements for multiple copies of documents

From 1 October 2016 it will no longer be necessary to provide duplicate copies of Patents Form 51. This form is used to appoint or change a patent agent.

It will also no longer be necessary to file an international patent application in triplicate, when using the IPO as a receiving office under the Patent Cooperation Treaty.

Changes taking place on 6 April 2017

Allowability of omnibus claims

Patent claims define the scope of the monopoly provided by the patent by setting out the technical features present in the invention. Sometimes claims are drafted in such manner as to refer generally to the description or drawings included in the patent application. Such claims do not state the technical features of the invention claimed and are known as "omnibus claims". Understanding these claims can be difficult and can lead to a lack of clarity about the scope of protection provided by the patent. This can make it difficult for businesses to know if they are likely to infringe a competitor's patent.

From 6 April 2017 it will generally no longer be allowable to include omnibus claims in UK patent applications. They will only be allowed if the technical features of the invention cannot otherwise be clearly and concisely defined using words, a mathematical or chemical formula or any other written means. From 6 April 2017 onwards, patent examiners will raise objections to any omnibus claims present in pending patent applications. But they will not do so if the patent application has a compliance period which expires before 6 April 2017.

Furthermore, from 6 April 2017 it will no longer be possible to amend a granted patent to insert an omnibus claim. However the presence of an omnibus claim in a granted patent will not be a ground for revocation. Any granted patents including such claims will remain valid.

Limiting the use of omnibus claims will increase legal certainty for businesses looking to determine whether they can operate in a particular technology area. It will bring UK requirements into line with the requirements of the European Patent Convention (EPC) and the international Patent Cooperation Treaty (PCT).

Address used to receive renewal reminders

Currently, patent holders must notify the IPO annually of the address to which renewal payment reminders are sent (where it differs from the registered address for service). From 6 April 2017, it will no longer be necessary to make this annual notification. Instead, the IPO will continue to use the address previously provided, until it is cancelled or updated by the patent holder. This change will reduce an administrative burden for patent holders. It will also provide them with an increased opportunity to use our online renewal services. These services are not available if an address is being notified at the same time as payment of the renewal fee.

****

Ch. 3

Trade Marks-Obtaining Protection

What is a Trademark?

A trademark is a symbol or a sign placed on, or used in relation to, one trader's goods or services to distinguish them from similar goods or services supplied by other traders. Section 1 of the Trade Marks Act 1994, which is the main legislation covering trade marks, defines a trade mark as any sign capable of being represented graphically which distinguishes the goods or services of one business from those of another.

The Intellectual Property Office (IPO) covers all matters relating to Trade Marks.

The vast majority of goods and services are covered by 'ordinary' trade marks. These marks function to indicate the trade origin, in other words they link the owner of the mark to the goods or services, and the goods or services to the owner. See below for more about trade mark definitions.

However, there are certain marks that do not have the same function as a ordinary trade mark. They are called Certification marks or Collective marks.

Certification marks

A certification mark is a specific type of mark. They provide a guarantee that the goods or services bearing the mark meet a certain defined standard or possess a particular characteristic.

The owner of the mark will define those standards or characteristics. Such marks are usually registered in the name of trade associations, government departments, technical institutes or similar bodies.

Collective marks

A collective mark is a specific type of trade mark which indicates that the goods or services bearing the mark originate from members of a trade association, rather than just one trader.

Domain names

A domain name is a name by which a company or organization is known on the Internet. It usually incorporates the company name, or other identifier.

To register a domain name you must apply to an accredited Registrar. A list of accredited and accreditation-qualified Registrars can be found on the ICANN (Internet Corporation for Assigned Names and Numbers) web site.

Very importantly, being the owner of a registered trade mark, does not automatically entitle you to use that mark as a domain name.

The main reason being, that the same trade mark can be registered for different goods or services and by different proprietors. Also, someone may have already, and quite legitimately, registered the domain name, perhaps with its use being connected with unregistered goods or services.

The opposite also applies, if your domain name has been properly registered, it does not automatically follow that a similar trade mark will satisfy the requirements for trade mark registration, and/or it may be confusingly similar to someone else's earlier trade mark.If you feel that a domain name has been registered unlawfully or maliciously then you should take appropriate legal advice. Alternatively, you can get advice from Nominet, www.nominet.org.uk who also offer a Dispute Resolution Service.

Historical background

Traders have, from the earliest times, distinguished their goods by marking them. By the 19th century it had become very clear that marks applied to goods that had become distinctive had an intrinsic value and needed some form of legal protection lacking at the time. Such protection was available through the use of Royal Charters and court action, which involved injunctions or action for infringement, although clearly this was not adequate or far reaching enough.

The Trademark Registration Act 1875 was passed to overcome the difficulties encountered in court actions. The Act established a statutory Register of Trademarks that is still in use today. The Register provides the trademark owner with proof of title to, and exclusive rights of use of, the trademark for the goods in respect of which it is registered. The Act of 1875 also laid down the essentials of a trademark. A number of Acts followed, the Patents, Designs and Trademarks Act 1883, the Trademarks Act 1905 and the Trademarks Act 1919. These Acts culminated in the 1938 Trademarks Act which in turn was replaced by the 1994 Trademarks Act.

International Provisions

There are a number of international conventions and arrangements that give some international recognition to national trademarks. These are the Paris Convention, The Madrid Arrangement and the Protocol to the Madrid arrangement (Madrid Protocol). There is also a Community Trademarks System that creates a trademark that gives rights throughout the European Community.

Definition of a trade mark

The 1994 Trademarks Act s.1(1) provides that a trade mark is a sign capable of being represented graphically, capable of distinguishing goods or services of one undertaking, from those of another undertaking. There are a number of elements in the definition:

a) *A 'sign'.* The concept of a sign in UK trademark law is very broad indeed. Although there is no clear definition, signs provided in the UK include works, designs and shapes and also more unconventional marks such as sounds and smells. A sign can be regarded as anything that conveys information

b) *Graphic representation.* Signs must be represented graphically, i.e. be represented in such a way that third parties may determine and understand what the sign is. This requirement is normally satisfied by including an image of the mark in the trade mark application. However, it has been suggested that provision of an image is not absolutely necessary provided that third parties can clearly identify the mark from the description It may be difficult to graphically represent unconventional marks, but practice dictates for example that sound marks are represented by music notation and that for shape marks it is best to submit line drawings or photographs. Applications for colour marks will usually include a representation of the colour and so on.

c) *Capable of distinguishing.* Signs must be capable of distinguishing goods or services of one undertaking from another undertaking. Any sign that has the capacity to distinguish will satisfy this requirement.

Signs not satisfying the s.1 (1) requirements

Signs which do not meet the definition of 'trade mark' provided in the Trademarks Act 1994 will not be registered. In addition, it is important for an applicant not to make a mistake as to the graphic representation as the opportunities to correct or amend are very limited. TMA 1994 s.39 prevents the correction of errors in a trade mark application that would substantially affect the identity of the trade mark. This is mitigated by the fact that it is IPO practice to examine marks for graphic representation before a filing date is allocated.

Signs must also be capable of distinguishing the goods or services of one undertaking from those of other undertakings. As noted above, this is not a high standard and, in effect, it will only bar those signs that are incapable of functioning as trademarks.

Marks devoid of distinctive character or those consisting of exclusively descriptive or generic signs are prohibited unless it can be shown that before the application was made, a mark has acquired a distinctive character as a result of a use made of it. This proviso to the TMA 1994 ss.3 (1)(b)(c) and (d) means that there is no absolute prohibition as a matter of law on non-distinctive, descriptive and generic marks. Such marks may be registered where they have become factually distinctive upon use despite the provisions stated in the TMA 1994 s.3 (1)(b)-(d). This proviso does not apply to TMA 1994 s.3(1)(a) or any other absolute ground for refusal.

Marks devoid of distinctive character

TMA 1994 s.3 (1)(b) prevents the registration of marks that are not, prima facie, distinctive. An example might include a surname common in the UK. Trademarks will only fail where they are not distinctive by nature and have not become distinctive by nature.

Signs that are exclusively descriptive

For a sign to be open to objection under TMA 1994 s.3 (1)(c) the trademark must consist exclusively of a sign which may be used in trade to describe characteristics of the goods or services. The sub-categories of TMA 1994 s.3 (1)(c) are:

1) Kind. Terms indicating kind or type that should be free for all traders to use, e.g. PERSONAL for computers, are not normally registrable.
2) Quality. Laudatory words, e.g. PERFECTION, are not usually registrable.

3) Quantity. The IPO gives the example that 454 would not be registrable for butter, as butter is frequently sold for domestic consumption in 454g (1lb) packs. Where numerical marks are not descriptive or otherwise objectionable, they may be registered.
4) Intended purpose. Generally, words referring to the purpose of goods or services are not registrable.
5) Value. Signs pertaining to the value of goods or services are not normally registrable, e.g. BUY ONE GET TWO FREE.
6) Geographical origin. Geographical names are not usually registrable unless used in specific circumstances.
7) Time of production of goods or the rendering of services. Typically, marks such as SAME DAY DELIVERY for courier services or AUTUMN 2004 for haute couture would not be registrable.
8) Other characteristics of goods and services. For example, a representation of the good or service would not usually be registrable.

Marks falling into any of these categories may still be registrable if they have become distinctive upon use.

Signs that are exclusively generic
TMA 1994 s3 (1)(d) prohibits the registration of signs or indications that have become customary in the current language or in the bone fide and established practices of the trade. An example can be found in JERYL LYNN Trademark (1999) where an application for JERYL LYNN for vaccines was refused as the mark described a strain of vaccine and was not distinctive of the applicant.

Shapes that cannot be registered
Traditionally in the UK, shapes were not registerable. One case highlighting this was Coca-Cola's trademark application (1986).

However, the TMA 1994 makes it very clear that the shapes of goods and their packaging are now registrable (TMA 1994 s.1 (1)), but the TMA 1994 s.3 (2) excludes certain shapes from registration. This is an area of trademark law that has lacked clarity. The TMA 1994 provides that the following shapes are not registrable:

1) Where the shape results from the nature of the goods themselves. Inherent shapes therefore cannot be registered
2) Where the shape of the goods is necessary to achieve a technical result (TMA 1994 s.3 (2)(b). Functional shapes are therefore not registrable.
3) Where the shapes gives substantial value to the goods. In Philips (1999) the Court of Appeal suggested that a valuable shape in this context can be identified where the shape itself adds substantial value, e.g. the shape adds value via eye appeal or functional effectiveness. In contrast, shapes that are valuable because they are 'good trademarks' would not fall foul of the TMA 1994.

Marks likely to give offence or deceive

A mark will not be registered if it is contrary to public policy or accepted principles of morality (TMA 1994 s.3 (3)(a) or is of such a nature that it is likely to deceive the public. For example, as to the nature, quality or origin of the goods or services.

Relatively few marks are deemed to be contrary to public policy or morality. Morality should be considered in the context of current thinking and only where a substantial number of persons would be offended should registration be refused.

Marks prohibited by UK or EC law

The registration of marks whose use would be illegal under UK or Community law is precluded by TMA 1994 s.3(1)(d).

Protected emblems

TMA 1994 s.4 provides details of marks that are considered to fall into the category of specially protected emblems, e.g. marks with Royal connotations, and the Olympic symbol cannot be registered. Marks containing such emblems cannot be registered without consent.

Applications made in bad faith

The key statute here is Section 3(3)(a) and (b) and section 3(6) Trade Marks Act 1994, Art 3(1)(f) and (2)(d) Directive on the Legal Protection of Trade Marks:

(3) A trade mark shall not be registered if it is -

(a) contrary to public policy or accepted principles of morality, or
(b) of such a nature as to deceive the public

(6) A trade mark shall not be registered if or to the extent that the application is made in bad faith.

There is no requirement that a mark needs be used prior to the application for registration, but the applicant must have a bona fide intention to use the mark and applications may be refused when they are made in bad faith. Therefore, so-called ghost applications should be caught by this section.

Relative grounds for refusal

Section 5(1) Trade Marks act 1994, Art 4(1)(a) Directive on the Legal Protection of Trade marks:

(1) A trade mark shall not be registered if it is identical with an earlier trade mark and the goods or services for which the trade

mark is applied for are identical with the goods or services for which then earlier trade mark is protected.

The applicant must also overcome the relative grounds for refusing registration. These relate to conflict with earlier marks or earlier rights. The 'earlier mark' (TMA 1994 s.6) might be a trademark registered in the UK or under the Madrid Protocol. Alternatively it might be a CTM or a well-known mark (the latter are entitled to protection as per article 6 of the Paris Convention for the Protection of Industrial Property 1883).

There is no provision for honest concurrent use in the TMA 1994. As it has been made clear that a trade mark application must be refused, irrespective of honest concurrent use, if the registered proprietor objects, this provision is of limited value to the applicant. If the proprietor of the registered mark objects, honest concurrent use provides no defence.

Conflict with an earlier mark for identical goods or services

The TMA 1994 s.5 (1) only provides the narrowest relative ground for refusing registration: a mark identical to an earlier trademark and used for identical goods and services will not be registered. The requirement of 'identical goods and services' is sufficiently broad in scope to include cases where the applicants mark is identical to only some of the goods and services for which the earlier mark is registered, but to 'constitute an 'identical mark' a very high level of identity between the marks is required.

The registration of similar marks for the same or similar services is only prohibited where confusion on the part of the public is likely to arise (TMA 1994 s.5 (2). Specifically what is prohibited is the registration of:

1) Identical marks for similar goods or services or
2) Similar marks for identical/similar goods or services where, because of the identity or similarity, there is a likelihood of

confusion on the part of the public, which includes the likelihood of association with the earlier trade mark.

What constitutes 'confusing similarity' has been considered at length. Confusion has to be appreciated globally taking into account all factors relevant to the case. These factors include:

- The recognition of the earlier trade mark on the market
- The association that can be made between the registered mark and the sign
- The degree of similarity between the mark and the sign and the goods and the services, the degree of similarity must be considered in deciding whether the similarity is sufficient so as to lead to a likelihood of confusion

It has also been made clear that 'likelihood of association' is not an alternative to 'likelihood of confusion" but serves to define its scope. This means that if the public merely makes an association between two trademarks, this would not in itself be sufficient for concluding that there would be a likelihood of confusion. There is no likelihood of confusion where the public would not believe that goods or services came from the same undertaking.

Conflict with a mark of repute

A mark that is identical or similar to an earlier mark will be refused registration in respect of dissimilar goods or services where the earlier mark is a mark of repute and the use of the later mark would, without the cause, take unfair advantage of or be detrimental to the reputed mark's distinctiveness or reputation. (TMA 1994 s5 (3).

A mark of repute is a mark with a reputation in the UK (for CTM applications it must have a reputation in the EU). In deciding as to whether a trade mark has a reputation, the ECJ has provided some guidance (General Motors Corp v Yplon) (2000). Repute would be judged with reference to the general public or to a specific section of the public, and the mark must be known to a significant portion of that public.

Relevant indicators of the public's knowledge of the mark include the extent and duration of the trade marks use, its market share and the extent to which it has been promoted.

In order for registration to be refused under s.5 (3) use of the applicants mark will have to take unfair advantage of or be detrimental to the reputed marks distinctiveness or reputation. In OASIS STORES LTD's application (EVEREADY) (1998) it was said that merely being reminded of an opponents mark did not itself amount to taking unfair advantage. The fact that the applicant did not benefit to any significant extent from their opponent's reputation and the wide divergence between the parties goods was relevant, s.5 (3) could not be intended to prevent the registration of any mark identical or similar to a mark of repute.

Conflict with earlier rights

TMA 1994 s.5 (4) provides that where a mark conflicts with earlier rights, including passing off, design rights and copyright the mark will not be registered.

Surrender, revocation, invalidity, acquiescence and rectification

Surrender. It is possible to surrender a trademark with respect to some or all of the goods or services for which it is registered. Marks may be revoked (removed from the registry) on several grounds: non-use because the mark has become generic; or because the mark has become deceptive. A mark will be invalid if it breaches any of the absolute grounds for registration. Where the proprietor of an earlier trade mark or other right is aware of the use of a mark subsequently registered in the UK and has, for a continuous period of five years, taken no action regarding that use the proprietor is said to have acquiesced. Where this is the case, the proprietor of the earlier mark or right cannot rely on his right in applying for a declaration of invalidity or in opposing the use of the later mark, unless it is being used in bad faith. Anyone with sufficient interest can apply to rectify an error or omission in the register. Such a rectification must not relate to matters that relate to the validity of the trademark.

Benefits of registration

Having taken into account all of the above, if you are now ready to apply for a trade mark, and you are certain that your mark is distinctive and doesn't clash with anyone else's mark then it is time to commence the registration process.

Importantly, registering your trade mark gives you the exclusive right to use your mark for the goods and/or services that it covers in the United Kingdom (UK).

If you have a registered trade mark you can put the ® symbol next to it to warn others against using it. However, using this symbol for a trade mark that is not registered is an offence.

A registered trade mark:

- may put people off using your trade mark without your permission
- makes it much easier for you to take legal action against anyone who uses your trade mark without your permission
- allows Trading Standards Officers or Police to bring criminal charges against counterfeiters if they use your trade mark
- is your property, which means you can sell it, franchise it or let other people have a licence that allows them to use it.

Protecting registered trade marks

If you don't register your trade mark, you may still be able to take action if someone uses your mark without your permission, using the common law action of passing off.

To be successful in a passing off action, you must prove that:

- the mark is yours
- you have built up a reputation in the mark
- you have been harmed in some way by the other person's use of the mark.

It can be very difficult and expensive to prove a passing off action.

If you register your trade mark, it is easier to take legal action against infringement of your mark, rather than having to rely on passing off.

How to apply for a trade mark

STEP 1: Before you apply

Search existing trade marks to see if there are any similar trade marks already in use which may affect your application for registration.

Check to see if it is a trade mark the IPO can accept.

STEP 2: Choose goods and services

Research and provide descriptions of the goods and/or services you are going to use your trade mark on. The Intellectual Property Office uses a classification system which divides the goods and services into 45 classes. See below for a detailed description of classifications.

STEP 3: Apply online – From £170. Check fees before applying.

Choose from the following examination services:

- Standard examination (£170 when paid in full with one class of goods or service)

  - No refund if your trade mark cannot be registered

- Right start (£200 for one class of goods or services)

  - Pay 50% on application

  - No refund, but only pay the balance if you decide to continue with the application.

The online application takes about 10 to 15 minutes to complete and you can 'Save for later' at any point.

How the IPO classifies

The Nice Agreement for the International Classification of Goods and Services provides that there are thirty-four classes of goods and also nine classes of services. Any application for registration must stipulate which classes, or sub-classes, in which registration is sought. Multi-class applications are possible and it would, in theory, be possible to register a mark in respect of all forty two classes. However, this is very unlikely as applicants must have a bona fide intent to use the marks for the prescribed goods and services (TMA 1994 ss.3 (6) and 32 (3)).

When applying for your trade mark, you will need to provide a list of the goods and/or services on which you intend to use your trade mark.

The following list provides you with general information about the types of goods and services which belong to each class. The classification system is divided between goods and services **goods** are in classes 1 - 34 and **services** are in classes 35 - 45.

Please note that these lists (called Class Headings) do not include all goods or services in a particular class.

If you do not know which class(es) your goods or services are in, contact the IPO via their e-mail advice service.

Goods

Read down the first column to identify the class and across the row to find a description of what is covered by that class.

Class 1	Chemicals used in industry, science and photography, as well as in agriculture, horticulture and forestry; unprocessed artificial resins, unprocessed plastics; manures; fire extinguishing compositions; tempering and soldering preparations; chemical substances for preserving foodstuffs; tanning substances; adhesives used in industry; unprocessed plastics in the form of liquids, chips or granules.
Class 2	Paints, varnishes, lacquers; preservatives against rust and against deterioration of wood; colorants; mordants; raw natural resins; metals in foil and powder form for painters, decorators, printers and artists.
Class 3	Bleaching preparations and other substances for laundry use; cleaning, polishing, scouring and abrasive preparations;

Read down the first column to identify the class and across the row to find a description of what is covered by that class.

soaps; perfumery, essential oils, cosmetics, hair lotions; dentifrices.

Class 4	Industrial oils and greases; lubricants; dust absorbing, wetting and binding compositions; fuels and illuminants; candles and wicks for lighting; combustible fuels, electricity and scented candles.
Class 5	Pharmaceutical and veterinary preparations; sanitary preparations for medical purposes; dietetic food and substances adapted for medical or veterinary use, food for babies; dietary supplements for humans and animals; plasters, materials for dressings; material for stopping teeth, dental wax; disinfectants; preparations for destroying vermin; fungicides, herbicides.
Class 6	Common metals and their alloys; metal building materials; transportable buildings of metal; materials of metal for railway tracks; non-electric cables and wires of common metal; ironmongery, small items of metal hardware; pipes and tubes of metal; safes; goods of common metal not included in other classes; ores; unwrought and partly wrought common metals; metallic windows and doors; metallic framed conservatories.
Class 7	Machines and machine tools; motors and engines (except for land vehicles); machine coupling and transmission components (except for land vehicles); agricultural implements other than hand-operated; incubators for eggs; automatic vending machines.
Class 8	Hand tools and hand operated implements; cutlery; side arms; razors; electric razors and hair cutters.
Class	Scientific, nautical, surveying, photographic,

Read down the first column to identify the class and across the row to find a description of what is covered by that class.

9	cinematographic, optical, weighing, measuring, signalling, checking (supervision), life-saving and teaching apparatus and instruments; apparatus and instruments for conducting, switching, transforming, accumulating, regulating or controlling electricity; apparatus for recording, transmission or reproduction of sound or images; magnetic data carriers, recording discs; compact discs, DVDs and other digital recording media; mechanisms for coin-operated apparatus; cash registers, calculating machines, data processing equipment, computers; computer software; fire-extinguishing apparatus.
Class 10	Surgical, medical, dental and veterinary apparatus and instruments, artificial limbs, eyes and teeth; orthopaedic articles; suture materials; sex aids; massage apparatus; supportive bandages; furniture adapted for medical use.
Class 11	Apparatus for lighting, heating, steam generating, cooking, refrigerating, drying, ventilating, water supply and sanitary purposes; air conditioning apparatus; electric kettles; gas and electric cookers; vehicle lights and vehicle air conditioning units.
Class 12	Vehicles; apparatus for locomotion by land, air or water; wheelchairs; motors and engines for land vehicles; vehicle body parts and transmissions.
Class 13	Firearms; ammunition and projectiles, explosives; fireworks.
Class 14	Precious metals and their alloys; jewellery, costume jewellery, precious stones; horological and chronometric instruments, clocks and watches.
Class	Musical instruments; stands and cases adapted for musical

Read down the first column to identify the class and across the row to find a description of what is covered by that class.

15	instruments.
Class 16	Paper, cardboard and goods made from these materials, not included in other classes; printed matter; bookbinding material; photographs; stationery; adhesives for stationery or household purposes; artists' materials; paint brushes; typewriters and office requisites (except furniture); instructional and teaching material (except apparatus); plastic materials for packaging (not included in other classes); printers' type; printing blocks.
Class 17	Rubber, gutta-percha, gum, asbestos, mica and goods made from these materials; plastics in extruded form for use in manufacture; semi-finished plastics materials for use in further manufacture; stopping and insulating materials; flexible non-metallic pipes.
Class 18	Leather and imitations of leather; animal skins, hides; trunks and travelling bags; handbags, rucksacks, purses; umbrellas, parasols and walking sticks; whips, harness and saddlery; clothing for animals.
Class 19	Non-metallic building materials; non-metallic rigid pipes for building; asphalt, pitch and bitumen; non-metallic transportable buildings; non-metallic monuments; non-metallic framed conservatories, doors and windows.
Class 20	Furniture, mirrors, picture frames; articles made of wood, cork, reed, cane, wicker, horn, bone, ivory, whalebone, shell, amber, mother-of-pearl, meerschaum or plastic which are not included in other classes; garden furniture; pillows and cushions.
Class 21	Household or kitchen utensils and containers; combs and sponges; brushes; brush-making materials; articles for

Read down the first column to identify the class and across the row to find a description of what is covered by that class.

cleaning purposes; steel wool; articles made of ceramics, glass, porcelain or earthenware which are not included in other classes; electric and non-electric toothbrushes.

Class 22	Ropes, string, nets, tents, awnings, tarpaulins, sails, sacks for transporting bulk materials; padding and stuffing materials which are not made of rubber or plastics; raw fibrous textile materials.
Class 23	Yarns and threads, for textile use.
Class 24	Textiles and textile goods; bed and table covers; travellers' rugs, textiles for making articles of clothing; duvets; covers for pillows, cushions or duvets.
Class 25	Clothing, footwear, headgear.
Class 26	Lace and embroidery, ribbons and braid; buttons, hooks and eyes, pins and needles; artificial flowers.
Class 27	Carpets, rugs, mats and matting, linoleum and other materials for covering existing floors; wall hangings (non-textile); wallpaper.
Class 28	Games and playthings; playing cards; gymnastic and sporting articles; decorations for Christmas trees; childrens' toy bicycles.
Class 29	Meat, fish, poultry and game; meat extracts; preserved, dried and cooked fruits and vegetables; jellies, jams, compotes; eggs, milk and milk products; edible oils and fats; prepared meals; soups and potato crisps.
Class 30	Coffee, tea, cocoa, sugar, rice, tapioca, sago, artificial coffee; flour and preparations made from cereals, bread, pastry and

Read down the first column to identify the class and across the row to find a description of what is covered by that class.

confectionery, ices; honey, treacle; yeast, baking-powder; salt, mustard; vinegar, sauces (condiments); spices; ice; sandwiches; prepared meals; pizzas, pies and pasta dishes.

Class 31   Agricultural, horticultural and forestry products; live animals; fresh fruits and vegetables, seeds, natural plants and flowers; foodstuffs for animals; malt; food and beverages for animals.

Class 32   Beers; mineral and aerated waters; non-alcoholic drinks; fruit drinks and fruit juices; syrups for making beverages; shandy, de-alcoholised drinks, non-alcoholic beers and wines.

Class 33   Alcoholic wines; spirits and liqueurs; alcopops; alcoholic cocktails.

Class 34   Tobacco; smokers' articles; matches; lighters for smokers.

Services

Read down the first column to identify the class and across the row to find a description of what is covered by that class.

Class 35   Advertising; business management; business administration; office functions; electronic data storage; organisation, operation and supervision of loyalty and incentive schemes; advertising services provided via the Internet; production of television and radio advertisements; accountancy; auctioneering; trade fairs; opinion polling; data processing; provision of business information; retail services connected with the sale of [list specific goods].

Class 36   Insurance; financial services; real estate agency services; building society services; banking; stockbroking; financial

Read down the first column to identify the class and across the row to find a description of what is covered by that class.

	services provided via the Internet; issuing of tokens of value in relation to bonus and loyalty schemes; provision of financial information.
Class 37	Building construction; repair; installation services; installation, maintenance and repair of computer hardware; painting and decorating; cleaning services.
Class 38	Telecommunications services; chat room services; portal services; e-mail services; providing user access to the Internet; radio and television broadcasting.
Class 39	Transport; packaging and storage of goods; travel arrangement; distribution of electricity; travel information; provision of car parking facilities.
Class 40	Treatment of materials; development, duplicating and printing of photographs; generation of electricity.
Class 41	Education; providing of training; entertainment; sporting and cultural activities.
Class 42	Scientific and technological services and research and design relating thereto; industrial analysis and research services; design and development of computer hardware and software; computer programming; installation, maintenance and repair of computer software; computer consultancy services; design, drawing and commissioned writing for the compilation of web sites; creating, maintaining and hosting the web sites of others; design services.
Class 43	Services for providing food and drink; temporary accommodation; restaurant, bar and catering services; provision of holiday accommodation; booking and reservation services for restaurants and holiday

Read down the first column to identify the class and across the row to find a description of what is covered by that class.

accommodation; retirement home services; creche services.

Class 44	Medical services; veterinary services; hygienic and beauty care for human beings or animals; agriculture, horticulture and forestry services; dentistry services; medical analysis for the diagnosis and treatment of persons; pharmacy advice; garden design services.
Class 45	Legal services; conveyancing services; security services for the protection of property and individuals; social work services; consultancy services relating to health and safety; consultancy services relating to personal appearance; provision of personal tarot readings; dating services; funeral services and undertaking services; fire-fighting services; detective agency services.

After you apply

When the IPO receives your application, they check it to make sure it has all the information they need. If there are any problems with your form, they will contact you although they cannot refund your application fee or premium for any reason and you cannot alter your mark after you have sent in your application form. The IPO will capture your details onto their database. They then issue your filing receipt and application number within a few business days:

- by email for e-filed applications
- by post for paper filed applications

Examining your mark

When the application is complete and the appropriate fee has been paid, it will be sent to an examination team who will examine your mark.

Options following an objection
Depending on the objection raised, you have a variety of options in dealing with the objection. If you need more time to complete certain actions, you can apply for an Extension of Time.

What happens once my mark is accepted?
Once the IPO has accepted your trade mark they publish it in their weekly Trade Marks Journal and write to tell you the publication date and the number of that Journal.

Is your mark acceptable?
The examination of your application will decide whether it meets the requirements of the Trade Marks Act 1994 and Trade Mark Rules 2000 (as modified).

Earlier potentially conflicting trade marks
The IPO will search their register to identify earlier potentially conflicting trade marks. These are marks which may be the same or similar to your mark and for the same or similar goods or services

Your examination report
The IPO normally issues their examination report within around 30 days of receiving your application. In this report they list their objections and requirements and tell you how long you have to reply. You will have the opportunity to persuade the examiner that the objection(s) are not valid or to make amendments to your specification of goods and services. If the only problem with your mark is earlier rights, you only get one chance to discuss this with the IPO.

Publishing your mark
When your application has overcome any objections and you are happy to proceed, the IPO will publish your application in their

Trade Mark Journal where your application is open to opposition for a 2 month period by any of the earlier marks they have notified, or any other party. This period can be extended to 3 months by anyone considering opposition. Opposition could result in your mark not being registered and you having to pay costs. If your mark is not opposed, it will become registered and you are free to use and enforce your trade mark.

Inspecting documents

The IPO will tell you how your trade mark application is progressing. You can also check the status of your trade mark or trade mark application on their official register.

Check status

You can enter your trade mark number into the IPO number search to check the status of your trade mark or trade mark application.

Inspection of documents held in the Registry

You may inspect the register of trade mark documents filed at the registry after 31 October 1994 in relation to:

- an application filed on/after that date which has been published, or
- a registered mark.

However, some documents may be excluded from this provision under the terms of Rule 50, which provides a full list of excluded documents. The provisions in Rule 50 take precedence over the disclosure provisions of the Freedom of Information Act 2000, as they are one of the exemptions under section 44 of that Act.

To view documents

You should Contact the IPO and give notice of your visit to allow files to be retrieved from their stores. There is a handling charge of

£5 per file which includes the cost of any prints taken, though the IPO can quote for their actual costs where particularly large files are involved.

To obtain copies of documents

If you are unable to visit the Office, you may request copies of the documents held on IPO files (subject to the inspection rights shown above), from Sales at the following address:

Intellectual Property Office
Concept House
Cardiff Road
Newport
South Wales
NP10 8QQ
Telephone 0300 300 2000

E-mail Please ensure that your full postal address and daytime telephone number are included in your e-mail.

The cost is £5 per file copied, though again the IPO reserves the right to quote for their actual costs where particularly large files are involved

Trade mark forms and fees

The following trade mark forms can be filed online:

- Trade mark application (TM3)
- Trade mark renewal
- Notice of threatened opposition

For a list of current fees relating to filing, both paper and online you should visit the IPO website.

Managing Trade Marks

Renewing your trade mark and changing information

There are limited ways in which you can change the details of your trade mark and you must renew your trade mark every 10 years to keep it in force. You should be be aware that companies and individuals are sending misleading invoices to applicants and registered owners of trade marks. These communications may appear to be official but they are not linked to the Intellecual Property Office or any other Government organisation.

Renewing your trade mark

You can renew your trade mark for ever. However, to keep your trade mark in force, you must renew it on the 10th anniversary of the filing date and every 10 years after that.

Restoration

If you do not renew your trade mark in the 6 months after the renewal date, you have a further 6 months to apply to restore your registration.

Changing

Transferring ownership (assigning)

You can transfer ownership of your trade mark to someone else. This includes any change of ownership as a result of company mergers.

Transferring ownership (assigning) Community and Madrid trade marks

You can only transfer ownership of your Madrid or Community trade mark through OHIM or WIPO. This includes any change of ownership as a result of company mergers.

Appointment or change of representative

You can appoint or change a trade mark attorney at any time by completing the appropriate form.

Change of owner's name, addres or e mail address

If you change your name, address or email address, you should tell the IPO so that they can update the trade marks register.

Correcting an error

You may be able correct an obvious error in your application or trade mark registration.

Part surrender of a trade mark

You can use the form TM23 to surrender some of the rights in your mark such as some of the goods or services registered. Once surrendered the rights cannot be reinstated.

Cancelling your trade mark

You can cancel your trade mark registration at any time.

Exploiting your trade marks

Licensing

A licence gives you permission to use someone else's trade marks. The terms of the licence are between you and the licensor and the IPO do not have any powers to investigate the validity of any licence you might agree.

When your licence agreement ends or the licensees details are changed you should use the Form TM51so that the IPO can record this on the register. There is a fee of £50 per form.

Mortgaging

You can use your trade mark as security for a loan. The mortgagor has a legal right in your trade mark until you repay the loan. You or your mortgagor should register the mortgage (security interest) with the IPO on form TM24 and they will then record it in the register. When it has been repaid they can cancel the details from the register.

You can cancel the 'security interest' from the register by filing form TM24C.

You can also record a financial interest in someone else's registered trade mark.

There is a fee of £50 per form.

Marketing

You may want to involve others to help exploit, develop or marketing your trade mark.

Certified and uncertified copies

The IPO will supply a copy of a trade mark or trade mark application, upon request and payment of the fee. You can also access the register on-line and get trade marks information at no cost.

Certified copies give proof that the IPO issued them.

Uncertified copies are photocopied documents.

What type of copy do I need?

You must use certified copies:

- when applying for a trade mark abroad
- when needed as evidence in a court of law, for example, if you are involved in legal action to enforce or defend your trade mark rights.

You can use **uncertified** copies for your personal reference or research.

Costs and how to pay

- The fee for a Certificate of the Registrar is £20
- The fee for a photocopy (uncertified) is £5

You must send a completed form FS2 (fee sheet), with this form.

To request a certified copy

- Use form TM31R Request for a certified copy.
- Use a separate form for each trade mark or trade mark application.

To request an uncertified copy:

- E-mail the IPO
- Phone them on 0300 300 2000
- Fax them on + 44 (0)1633 817777.

Send the IPO your request

Send your form, fee sheet and fee to:
Intellectual Property Office
Concept House
Cardiff Road

Newport
South Wales
NP10 8QQ

You must apply to the The Office for Harmonisation in the Internal Market to certify community trade marks.

You must apply to the World Intellectual Property Organisation (WIPO) to certify international trade marks

Transferring or selling the ownership of a trade mark

If you transfer or sell the ownership of your trade mark or the ownership changes following company mergers, then you must tell the IPO by using the form TM16 so that they can record it on the register of trade marks. Assignment is the legal term for transferring ownership and the form TM16 is the form the IPO use to record this. It does not replace a formal assignment document. There is a £50 fee per form.

The form must be signed by you and the new proprietors. If this is not possible, then documentary evidence of the transfer of ownership must be provided.

Send your request to the Intellectual Property Office, address as above.

What happens next?

The IPO will record the assignment in the register of trade marks and confirm this in writing

Enforcing your trade mark-Using the ° Symbol

You do not have to identify your trade mark as registered but you can use the ° symbol or the abbreviation "RTM" (for Registered Trade Mark) to show that your trade mark is registered, the mark can be registered somewhere other than in the United Kingdom.

The ® symbol is usually placed on the right-hand side of the trade mark, in a smaller type size than the mark itself, and in a raised (superscript) position; none of this is compulsory. If you do not have the ® symbol available, you can use the abbreviation "RTM".
You would break the law (Section 95 of the Trade Marks Act 1994) if you use the registered symbol ® or the abbreviation "RTM", on a mark that is not registered anywhere in the world.

Am I breaking the law by using "TM" on my trade mark?

No, as this does not indicate that your trade mark is actually registered, only that it is being used as a trade mark. The symbol 'TM' has no legal significance in the United Kingdom.

Resolving disputes

The IPO always encourages parties who are in dispute to resolve their differences before seeking a judgment by the office.

Before proceedings commence

Lord Woolf's 1996 report 'Access to Justice' identified the need for parties to see legal action as a last resort. He suggested that they should first try to settle matters outside the judicial system. These principles are reflected in the Civil Procedure Rules which were introduced in April 1999. In line with those Rules, if an action is launched before the Registrar and there is no prior contact between the parties, they may be penalised when the costs of the case are determined. So if you are thinking of taking legal action you should attempt to resolve the matter before launching any action.

Requests for stays or suspensions in inter partes proceedings

Where a stay or suspension is requested because the parties are trying to negotiate an amicable settlement, the parties will need to show what they have already done to resolve the dispute.

If the IPO is not satisfied that those negotiations are making progress they will not allow any further extensions to the stay of proceedings.

Hearing or written decision

When any period allowed for the filing of evidence is over the IPO will offer the parties either a hearing or a decision from the papers already filed. In either case the decision will resolve the dispute. The decision will be open to appeal.

Mediation

Mediation is another route that the IPO will be actively encouraging. It is another way that parties can resolve their dispute.

Cooling-off period

Gives both sides in potential opposition proceedings the chance to agree to settle their differences within a cooling-off period, without going through the full legal procedure.

The Company Names Tribunal

The Company Names Tribunal adjudicate on opportunistic company name registrations.

Counterfeiting of trade marks

If someone deliberately uses your registered trade mark, without your knowledge or consent, they may be guilty of the crime of counterfeiting.If there is sufficient evidence, the Police or Trading Standards Officers can take criminal proceedings under trade marks law. The IPO is not responsible for policing the Register of trade marks, and cannot advise you about what legal action to take to protect your mark.

If you suspect that someone is passing off, infringing, or counterfeiting goods or services under your mark, it is recommended that you consult your local Trading Standards Office. You can also visit the official trading standards website for more information. You can also seek appropriate professional help, from a solicitor or a trade marks attorney.

European & International Trade mark's

If you want to use your trade mark in countries other than the United Kingdom, you can apply directly to the Trade Mark Office in each country.

You can use a single application system to apply for an *International trade mark* (for certain countries throughout the world), or a *Community trade mark*. (for protection in Europe)

Both these single application systems cover many countries including the United Kingdom and offer a number of other potential benefits, including:

- less to pay;
- less paperwork;
- lower agents' costs;
- faster results;
- easy application

The International route

You can apply to register your trade mark through the international route in countries which are party to the Madrid Protocol through the World Intellectual Property Organisation (WIPO). Currently more than 70 countries are members, including the United States of America, Australia and members of the European Union (EU).

The European route

You can apply for a European Community trade mark through the European route via the Office for Harmonization in the Internal

Market (trade marks and designs) (OHIM). The Community trade mark gives protection in all European Union (EU) countries.

Re-registration of UK marks in other countries

The IPO provides a list of web links that will direct you to their most up-to-date lists of the countries in which trade mark protection may be extended. You can access this list via the Professional Section of their website.

Using and buying trade marks

You may be able to buy or use other people's trade marks. If you want to use other people's trade marks, you usually need permission. If you use registered trade marks without permission, you are infringing the trade mark and the owner can take legal action against you and claim damages. If you want to use a registered trade mark, you can approach the owner to agree a licence with them. You may also be able to buy the trade marks rights from the owner. This results in transferring the ownership, or assigning it, to you.

Infringing

If you use an identical or similar and confusable trade mark for identical or similar goods or services to a trade mark already in use - you are likely to be infringing the earlier mark. You can search IPO records to find the owner of a registered trade mark.

Registrable transactions

You can use certified copies to prove details about a trade mark or application. You can use uncertified copies for your personal reference or research.

Buying
Transferring ownership

If you buy a trade mark, you must tell the IPO that you are the new owner.

Infringement-What is trade mark infringement?

If you use an identical or similar trade mark for identical or similar goods and services to a registered trade mark - you may be infringing the registered mark if your use creates a likelihood of confusion on the part of the public. This includes the case where because of the similarities between the marks the public are led to the mistaken belief that the trade marks, although different, identify the goods or services of one and the same trader.

Where the registered mark has a significant reputation, infringement may also arise from the use of the same or a similar mark which, although not causing confusion, damages or takes unfair advantage of the reputation of the registered mark. This can occasionally arise from the use of the same or similar mark for goods or services which are dissimilar to those covered by the registration of the registered mark.

What about unregistered trade marks?

There is no available remedy for trade mark infringement if the earlier trade mark is unregistered. Some unregistered trade marks may be protected under Common Law and this is known as Passing off. However, whether or not they are protected will depend on the particular circumstances, in particular:

- Whether, and to what extent, the owner of the unregistered trade mark was trading under the name at the date of commencement of the use of the later mark;
- Whether the two marks are sufficiently similar, having regard to their fields of trade, so as to be likely to confuse

and deceive (whether or not intentionally) a substantial number of persons into thinking that the junior user's goods and services are those of the senior user;

- The extent of the damage that such confusion would cause to the goodwill in the senior user's business.

I think that I maybe infringing, what should I do?

Get legal advice. There may be a number of potential courses of action or defences open to you, but this will very much depend on the particular circumstances of your case.

Some traders who think they may be infringing an earlier trade mark choose to cease trading under the offending sign, others choose to approach the earlier trade mark owner and attempt to negotiate a way forward that suits both parties, which may include a co-existence agreement.

If you decide that you are not infringing, or you have a good defence, you may decide to stand your ground or even to sue the trade mark holder for making unjustified threats. In the worst case scenario, you may have to change your trade mark and re-brand your products or services.

I think that someone else maybe infringing, what should I do?

Get legal advice as the most suitable course of action will depend on the particular circumstances of your case.

One potential option open to you is to write to the infringer. However you must be satisfied that the earlier trade mark that you own and the activities of the infringer justify this. This is because the law also protects traders from unjustifiable threats of trade mark infringement.

You may be able to negotiate a settlement which suits both parties, which may involve a co-existence agreement. Another option is that you may be able to get a court order to force the infringer to cease trading and pay compensation for damages.

However, infringement actions must be taken to the High Court or in Scotland, the Court of Session.

You can get legal advice, from The Institute of Trade Mark Attorneys (ITMA), or the Chartered Institute of Patent Attorneys (CIPA) or the Law Society..

****

Ch. 4

Registered Designs

Design

A design may be protected in a number of ways, in particular by the Community Design, (Registered Design and Unregistered Design) the UK Registered Design and the UK Unregistered Design Right. The UK Registered and unregistered design rights have been amended by the introduction of the Intellectual Property Act 2014 which came into force between October 2014 and the end of 2015. This is outlined at the end of this chapter.

Community Design

There are two forms of Community design, one subject to registration (RCD) and the other informal (UCD). The basic requirements are both the same, apart from the date at which novelty and individual character is tested.

A community design has a unitary character and has equal effect throughout the Community. It may only be registered, transferred, surrendered, declared invalid or its use prohibited in relation to the entire community.

The main legislation dealing with community design is Article 3 Community Design regulation OJ 2002 L341:

a) 'design' means the appearance of the whole or part of a product resulting from the features of, in particular, the lines, contours, colours, shape, texture and/or materials of the product itself and /or its ornamentation;

b) 'product means any industrial or handicraft item, including *inter alia* parts intended to be assembled into a complex product, packaging, get up, graphic, symbols and typographic typefaces but excluding computer programs:

c) 'complex product' means a product which is composed of multiple components which can be replaced, permitting disassembly and re-assembly of the product.

Article 4 (1) Community Design regulation

A design shall be protected by a community design to the extent that it is new and has individual character.

Novelty and Individual character

A design is new if no identical design (including a design with features which differ only in immaterial details) has been made available to the public. There is a proviso to this and that is a pre-existing design will be disregarded if it could not reasonably have become known in the normal course of business to the circles specialized in the sector concerned operating in the community.

A design has an individual character if the overall impression it produces on an informed user differs from the overall impression produced on such a user by any design which has been made available to the public.

A key case concerning novelty and individual character is that of Green Lane Products Ltd v PMS International Group Ltd (2008). In this case, a challenge to the validity of the claimant's Community design for spiky laundry balls was based on the defendant's similar shaped spiky balls used for massaging the human body.

It was established in this case that the prior art is not limited to the particular product for which the design was registered, as the scope of infringement is not limited to the product for which it was intended to apply the design. For example, the registration of a design intended for motor cars would protect also against its use for toys. The 'informed user' is not the same as the average consumer of trade mark law. The informed user has experience of similar

78

products and will be reasonably discriminatory and able to appreciate sufficient detail to decide whether or not the design under consideration
creates a different overall impression. The degree of design freedom is taken into account.

A key case concerning individual character and design freedom is that of Pepsico Inc's design (No ICD000000172) OHIM.

The design in question was for a disk having annular rings or corrugations applied to a promotional item for games. There was a challenge to the validity of the design. The design was declared invalid.

The legal principle was that the informed consumer would be familiar with promotional items and would pay more attention to graphical elements rather than minor variations in shape. Furthermore, although there were some constraints to design freedom, these were to do with cost and safety, and, otherwise, there was ample design freedom. Thus, the informed user may focus on certain aspects of a design and design freedom should be looked at in the round and some constraints may be present without significantly reducing the overall design freedom.

Time periods for testing novelty and individual character

The time when a design has been made available to the public differs between the RCD and the UCD (Registered and Unregistered). The RCD relevant date is the date of filing the application, or earlier priority date if there is one. The UCD relevant date is the date the design itself is first made available to the public.

There is a 12-month period of grace for the RCD so, for example the designer may market products relating to that design during that period before filing the application to register. "Under the bonnet' parts which are not seen during normal use of a complex product are not considered to be novel or have individual character.

Exclusions from Community Design

The below are exclusions from community design:

- Features dictated by technical function
- 'Must-fit' features (except in respect of modular systems which are protectable in principle)
- Designs contrary to public policy or morality
- ''Must-match' spare parts used to restore the original appearance of a complex product.

Duration

Registered Community Designs – five years from the date of filing. It may then be renewed for further periods of five years up to a maximum of 25 years. Unregistered Community Designs-three years from the date the design was first made available to the public.

For the purpose of the unregistered design in determining the start of the three years, it is made available to the public when it is published, exhibited, used in trade or other wise disclosed in such a way that, in the normal course of the business, these events could have reasonably have become known to the circles specialized in the sector concerned in the community.

Protection and infringement of a community design

The scope of protection of a community design resembles the test for individual character in that it is a question of whether the alleged infringing design, from the perspective of the informed user, does not produce a different overall impression compared with the protected design. Design freedom is taken into consideration.

The main legislation concerning protection of community design is Article 10 Community Design Regulation OJ 2002 L341:

1) The scope of protection conferred by a community design shall include any design which does not produce on the informed user a different overall impression.

2) In assessing the scope of protection, the degree of freedom of the designer in developing his design shall be taken into consideration.

The registered community design gives the rightholder a monopoly right which is infringed by a person using it without the rightholders consent. Use, in particular, includes making, offering, putting on the market, importing, exporting or using a product in which the design is incorporated or applied, or stocking such a product for those purposes.

For the unregistered Community design, it is required that the use in question results from copying the protected design. This also applies during the period of deferred publication where the design is registered but publication has been deferred. An applicant to register a Community design can defer publication by up to 30 months from the filing date, hence delaying the payment of the publication fee.

A key case concerning infringement was that of Procter and Gamble Co v Reckitt Benckiser (UK) Ltd (2008) which was a case on the alleged infringement of a registered Community design applied to a spray container for air fresheners.

It was found in this case that a design did not have to be clearly different, it was sufficient if it differed in a way that the informed user was able to discriminate. An initial decision that there had been an infringement was reversed.

Limitations on the rights to a Community design

The rights to a Community design (registered or unregistered) do not extend to the following acts:

- Acts done privately and for non-commercial purposes.

- Acts done for experimental purposes
- Reproduction for citation or teaching in accordance with fair practices without unduly prejudicing the normal exploitation of the design, providing the source is mentioned.
- Acts in respect of the repair of ships or aircraft temporarily in the Community.

UK registered design

The UK registered design has been modified as a result of the EU Directive harmonising registered design law throughout the European Community. As a result, the main principles in the EU and UK are virtually identical.

The main legislation dealing with registered designs in the UK is section 24A (2) of the Registered Designs Act 1949 and Regulation 1A (2) Community Design regulations 2005/2339, as amended. In addition, the Intellectual Property Act 2014 has made changes to both UK registered and unregistered designs as outlined below.

UK unregistered design right

The UK unregistered design right was introduced by the Copyright, Designs and Patents Act 1988 in an attempt to overcome the problems of protection of functional designs by means of copyright in drawings showing the designs, as highlighted in *British Leyland Motor Corp v Armstrong patents Co Ltd (1986).*

The main legislation Section 213(1), (2) and (4) Copyright, Designs and Patents Act 1988, as amended, states:

(1) Design right is a property right which subsists in accordance with this part in an original design.

(2) In this part, 'design' means the design of any aspect of the shape or configuration (whether internal or external) of the whole or part of an article.

(3) A design is not 'original' for the purposes of this part if it is commonplace in the design field in question at the time of its creation.

Exceptions to subsistence of UK unregistered design right
Sections 213 (3) Copyright, Designs and Patents act 1988 states:

(3) Design right does not subsist in -

(a) a method of principle of construction,

(b) features of shape or configuration of an article which -

(i) enable the article to be connected to, or placed in, around or against another article so that either article may perform its function, or

(ii) are dependant upon the appearance of another article of which the article is intended by the designer to form an integral part, or

(c) surface decoration.

The first exception, methods or principles of construction, is unlikely to be relevant in the majority of cases. then other exceptions are often referred to as the 'must fit' or 'must match' exceptions. Surface decoration is also excepted.

A key case in the area is that of Dyson Ltd v Qualtex (UK) Ltd (2006) which concerned various aspects of design right including the scope of the 'must fit' and 'must match' and surface decoration exclusions.

The facts of the case were that the defendant supplied duplicate spare parts (pattern parts) for the claimant's vacuum cleaners. The claimant sued on the basis of the unregistered design rights subsisting in the design of the parts of its vacuum cleaners.

The court found that the 'must fit' exclusion does not mean that the articles have to physically touch. A clearance between them, if it allows one article to perform its function, may be within the exclusion. the exclusion may apply where the two articles are designed sequentially one after the other.

For 'must match' exclusion it is the design dependency which is important. the more room there is for design freedom, the less likely the exception will apply. The reason for the surface decoration exclusion was because it was protected by copyright. Surface decoration could be applied to a two-dimensional article or three dimensional article or to a flat surface of a three dimensional article. Surface decoration was not limited to something applied to an existing article and it could come into existence with the surface itself. Surface decoration could be three dimensional such as beading applied to furniture. However, a feature having a function, such as ribbing on the handle of a vacuum cleaner, was unlikely to be surface decoration.

The Intellectual Property Act 2014

The Act seeks to modernise and simplify certain aspects of intellectual property law including the law relating to designs, in order to promote innovation.

What are the purposes of the changes to design rights?

The purpose of the design right changes are to:

- simplify design law and allow the intellectual property framework to better support innovation
- improve the enforcement of designs and understanding the design rights of others, and
- improve the processes associated with the design framework

What changes are made to design right protection?

Unregistered Designs

- Designer is now the first owner: One fundamental change to existing law is that the act changes the deemed first owner of unregistered designs so that unless otherwise agreed, the designer will be the owner of the designs and not the person who commissioned the designs. Historically the first owner has been the commissioner. This amendment will bring design law into line with UK copyright law
- Trivial design features limited: The definition of Unregistered Design Right has been amended to limit the protection for trivial features of a design
- Clarification of definition of 'design': The act clarifies the definition of design so that to be original a design must not be commonplace in a 'qualifying country' rather than in the 'relevant design field', which caused confusion as to its geographical coverage. Qualifying Country is defined in the Copyright Designs and Patents Act 1988 (CDPA)
- Qualifying Person amended: The act has amended the provisions related to qualifying persons who can claim unregistered design right so that those who are economically active in the EU and other Qualifying Countries (as set out in the CDPA) have protection
- New exceptions to infringement: The act extends the exceptions for infringing unregistered designs, so that acts done privately for no commercial purpose or for teaching will not infringe unregistered design rights. A similar exception applies for acts done for experimental purposes, this is to encourage innovation

Registered Designs

- Designer is now the first owner: The initial ownership position in respect of registered designs is also changed, as for unregistered designs i.e. the designer will be the initial owner unless otherwise agreed, rather than the commissioner.

- New Criminal Offence: The act makes intentional copying of a registered design a criminal offence. This applies to acts which take place in the course of business and the penalties for such an offence are now a fine or prison sentence. Again, this brings the penalties into line with sanctions for trade marks and copyright infringement.

- New exceptions to infringement: The act expands the exception from copyright infringement already available to registered UK designs to registered community designs, i.e. so that an authorised user of a UK or registered community design cannot be sued for infringement of associated copyright.

- Applicant does not need to be the owner of the design: The act removes the requirement for the applicant of a registered design application to be the proprietor of the design.

- New good faith exception to infringement: The act introduces a right of prior use, allowing a third party who has acted in good faith to continue to use a registered design which is subsequently registered by another. The aim of this amendment is to provide an entitlement to limited exploitation in respect of uses already made.

- New powers of enforcement: The act gives Trading Standards officers similar powers of enforcement for design offences as those already available to them in respect of copyright and trade marks.

- Simplified international registration: The act grants power for the Secretary of State to implement the Geneva Act of the Hague Agreement in the UK, this means that international registration procedures will be available for UK registered designs. At present UK designers can only access the Hague registration process via the EU community design registration.

- Harmonisation of financial liability: The act aligns the financial liability provisions for innocent infringement with those provisions under the Community Design Rights legislation.

- Simplified appeal process: The act allows a new route of appeal against Intellectual Property Office (IPO) decisions via an Appointed Person instead of appealing via the courts, such root already exist for trade mark appeals. This amendment is intended to allow appeals be cheaper and less time consuming.

- New opinions service for ownership, validity and infringement issues: The act provides for a voluntary non-binding opinion service to be introduced by the IPO which is similar to the opinions service which currently exists in respect of patents.

- Clarification: The act clarifies that proceedings for an offence committed against a partnership must be brought against the partnership.

- derived from a continuing programme of research, a report of which is intended for future publication, where disclosure would prejudice the report.

****

Ch. 5

How to Apply for a Registered Design

You can apply for Registered Design protection by filling out the appropriate application form(s), found on the IPO website, and paying the relevant fee.

Design application form DF2A

This is the most commonly used form. It is essential that you refer to DF2A notes and the 'How to apply to register a design' booklet for instructions on how to complete the form.

Surplus design application form DF2B

Use this form if the IPO has asked you to divide out a design from an earlier application.

Refer to DF2A and DF2B notes for guidance on how to complete the form, and also the 'How to apply to register a design' booklet (536Kb) for instructions on filling out the form.

You will need to send new copies of your representations for any designs on the DF2B *and* the design remaining on the original application.

Send the application

You should send your application to the IPO. You must include your completed DF2A and fee sheet forms, one copy of the illustrations of your design, and your fee. It is not advised to fax design applications as the loss of detail on your representations may affect the amount of protection you have, and you may lose your filing date if the IPO have to ask for better images.

How much does it cost?

If you wish to have your design or designs published and registered as soon as possible:

- It costs GBP £60 to apply to register a single design or the first design in any multiple application. For every additional design in any multiple application it costs GBP £40 per design.

If you wish to defer the publication of a design by up to 12 months:

- It costs GBP £40 to apply and to register a single design or the first design in any multiple application.For every additional design in a multiple application it costs GBP £20 per design.
- When you are ready to request publication and registration of deferred designs you will need to file Form DF2C and pay a publication fee of GBP £40 per design.

Why might I need to defer publication?

If you are applying for a patent or wish to delay the public disclosure of the design for any reason, you may wish to defer registration and publication of your design.

Changing or renewing your registered design

You can change or renew the details of your registered design in various ways.

Changing
Appoint or change an agent or contact details

You can appoint a specialist attorney, to deal with your application or some complicated part of the procedures for you.

Cancelling your design

You can cancel your own design registration at any time. The implications of cancellation are quite far-reaching.

Renewing-Renewing your design

To keep your registered design in force, you must renew it on the 5th anniversary of the registration date and every 5 years after that up to a total of 25 years.

Restoration

If you registered your design on or after 1 August 1989 and you do not renew it in the 6 months after the renewal date, you have a further 6 months, to apply to restore your registration.

Appointing or changing an agent or contact address

You can use DF1A to remove or appoint a person, company or agent to deal with your application or any other complicated part of our procedures for you. It can also be used to alter the contact details that the IPO hold for you. There is no fee for this.

What happens next?

The IPO will record the change or appointment and send written confirmation to you.

Correcting an error

You can correct an error in your design in the following ways:

- amending the details on your application form (except name and address); or
- amending errors in the illustrations of your design; or
- amending your entry in the register of designs

Errors can usually only be corrected following a clerical or obvious error. You will need to write to the IPO with your request. There is no fee for this. You should use the IPO Proprietor Search to make sure you have listed all the Designs affected. Send your request to:

What happens next?
The IPO will consider the request and, if acceptable, will record it on the Designs Register and send written confirmation to you.

Cancelling your registered design
You can cancel your design registration at any time. This action is usually taken

- to avoid legal action such as infringement proceedings; or
- because of an application to remove the registered design by a third party

You will need to make sure you are aware of any implications of cancelling your design, for example if you have licensed the design to someone based on your registration.

To cancel your own registration, fill in form DF19C There is no charge for this.

What happens next?
The IPO will cancel the registration and write to confirm that they have done so.

Renewing your design
To keep your registered design in force, you must renew it on the 5th anniversary of the registration date and every 5 years after that.

You can renew your design up for 25 years, but you may choose not to renew it or to voluntarily cancel it at any time. You can renew your registration up to 6 months before the renewal date. The IPO will write to remind you 3 months before renewal is due if you have not already renewed it.

How do I renew my registered design?

To extend the period of protection, fill in form **DF9A** and **FS2** and send the correct fee. The renewal fees are as follows (note that new fees will be introduced from october 2016):

Period	Fee
2nd	£130
3rd	£210
4th	£310
5th	£450

What happens next?

The IPO will write to confirm that they have renewed your design for the next 5-year period.

What if I don't renew in time?

If you do not pay your renewal fee by the renewal date your registration becomes expired. The IPO will write to confirm this, but you can still renew it within 1 month of the renewal date at no extra charge. You have a further 5 months in which to renew your design. However, there is a late payment fee of £24 for each of these months in addition to the renewal fee. If you do not renew your design within this 6 month period you can still apply to restore your design.

Restoring your design

After the 6-month late renewal period, there is a further 6 month period where you can apply to restore your design.

Restoration

If you registered your design on or after 1 August 1989 and you do not renew it in the 6 months after the renewal date, you have a further 6 months, to apply to restore your registration.

You cannot restore designs registered before 1 August 1989 that have not been renewed within the 6 months, allowed.

Fill in Form DF29

To restore a registered design, complete and submit Form DF29 enclosing the following:

- a statement of the reasons for your failure to renew, and
- a fee of GBP £120

What happens next?

the IPO will process your application to consider restoration of registered designs and give notice of the application in the Designs Journal. If they are satisfied that failure to pay the renewal fee was unintentional, they will write to tell you and ask you to send them:

- Form DF9A - Application to extent the period of protection and
- The correct renewal fee

Once they receive your forms and fees, they will restore your design to the register and publish that fact in the Designs Journal. If they are not satisfied that your failure to renew was unintentional, they will:

- write and explain why we have not accepted your reasons; and
- allow you 2 months to request a hearing.

If you do not request a hearing within 2 months, they will refuse your application and publish that fact in the DesigJournal.ns

Using and enforcing your design
You can decide how to use and enforce your registered design.

Using-Licensing and selling
You can give someone else permission to use your Registered Design by granting them a licence, or you can sell (assign) it.

Mortgaging
You can use your registered design as security for a loan. The mortgagor has a legal right in your design until you repay the loan.

Marketing
You may want to involve others to help exploit, develop or market your design.

Enforcing your design-Display your rights
There is no official symbol to show that a design is registered but you can display the design number on the product once it is registered.

Resolving disputes
The IPO always encourage parties who are in dispute to resolve their differences before seeking a judgment by them, but they do offer a mediation service which may help.

Certified and uncertified copies
The IPO will supply a copy of a design or design application, upon request and payment of a fee.

Licensing

You can give someone else permission to use your registered design by granting them a licence. The terms of any licence are entirely a matter between you and the licensee. You or the licensee should register the licence, or cancellation of a licence, with the IPO and they will record it in the register of designs.

Licensing of unregistered rights

The IPO offers some limited services in relation to licensing of unregistered Design Right.

Display your rights

There is no official symbol to show that a design is registered, but you can display the design number on the product once it is registered.

However, please note, under Section 35 of the Registered Designs Act 1949 as amended it is an offence to falsely claim that a design is registered.

Your design registration is not valid until the IPO have issued your certificate, so it is still an offence to claim your design is registered during the period of time in between applying for a registration and receiving confirmation from the IPO that registration has been granted. This includes applications which have been deemed acceptable but for which deferred registration and publication has been requested. During this period, you can state that design registration has been applied for, but not that it has been granted.

Resolving disputes

The IPO always encourages parties who are in dispute to resolve their differences before seeking a judgment by the office.

Before proceedings commence

Lord Woolf's 1986 report 'Access to Justice' identified the need for parties to see legal action as a last resort. He suggested that they should first try to settle matters outside the judicial system. This is the same approach as adopted with all disputes connected to Intellectual property. In the first instance, try to resolve the matter before action.

These principles are reflected in the Civil Procedure Rules which were introduced in April 1999. In line with those Rules, if an action is launched before the Registrar and there is no prior contact between the parties, they may be penalised when the costs of the case are determined. So if you are thinking of taking legal action you should attempt to resolve the matter before launching any action.

Mediation

Mediation is another route that the IPO will be actively encouraging. It is another way that parties can resolve their dispute.

Requests for stays or suspensions in inter partes proceedings

Where a stay or suspension is requested because the parties are trying to negotiate an amicable settlement, the parties will need to show what they have already done to resolve the dispute. If the IPO is not satisfied that those negotiations are making progress they will not allow any further extensions to the stay of proceedings.

Hearing or written decision

When any period allowed for the filing of evidence is over the IPO will offer the parties either a hearing or a decision from the papers already filed. In either case the IPO decision will resolve the dispute. The decision will be open to appeal.

Certified and uncertified copies

The IPO will supply a copy of a design or design application, upon request and payment of the relevant fee. You can also access the register on-line and get designs information free of charge.

You can request copies as filed, or as registered. Copies of applications which have not yet been registered may only be issued to the applicant or their agent.

- Certified copies (Certificates of the Registrar) give proof that the United Kingdom Designs Registry has received an application to register or has granted registration of a design, in accordance with the Registered Designs Act and Rules, and that that copy was formally issued by the Intellectual Property Office.
- Uncertified copies are photocopied documents.

What type of copy do I need?

You must use certified copies:

- When needed as evidence in a court of law, for example if you are involved in legal action to enforce or defend your design rights.

You may need to use certified copies:

- When applying for a design abroad and claiming a Priority Date from your earlier UK application.

You can use uncertified copies for your personal reference or research.

Costs and how to pay

- A Certificate of the Registrar costs £22.

98

- A photocopy (uncertified) costs £5 per file copied which includes postage, though the IPO reserve the right to quote for their actual costs where particularly large files are involved.
- A fee sheet must be filed with all fee bearing forms.

To request a certified copy
- Use form DF23 Request for a certified copy.
- Use a separate form for each design or design application.
- Send your completed form DF23, FS2 fee sheet and payment to the IPO

To request an uncertified copy

- E-mail the IPO at sales@ipo.gov.uk
- Phone the IPO on 01633 814184.
- Fax the IPO on 01633 817777.

Selling your design
You are free to sell or transfer ownership of your registered design to someone else. If you do, you must tell the IPO so they can record the change in the register of designs. This includes any change of ownership as a result of company mergers. The legal term for a transfer of ownership is an assignment.

Fill in Form DF12A
To tell the IPO you have sold your registered design you need to fill in form DF12A . There is no charge for this. Form DF12A is not a replacement for the assignment, merely the form that you should use to ask the IPO to record it.
You and the new owners or their representatives must sign the form. If any of them cannot do so, the IPO will accept other documents as proof that the assignment has taken place.

What happens next?

The IPO will record the details of the assignment in the register of designs and write to confirm that they have done so

Designs protection abroad

Registering your design in the UK does not protect it abroad.

If you want to register your design in countries other than the UK, there are a number of ways in which you can do so.

- You can apply for a Registered Community Design (RCD) covering the whole of the European Union ('EU').
- You can apply directly to most major countries of the world by making a separate application to each country in which you want protection.
- You can use the Hague System to apply to a number of different countries or territories at the same time, through a single application.

Using unregistered design rights

You may also be able to rely on automatic copyright and unregistered design rights in the countries concerned, eg. the Unregistered Community Design right (UCD) which covers the whole of the EU.

Claiming a priority date

If you are applying for a design in another country within 6 months of applying for the same design in the UK, or if you are applying in the UK within 6 months of filing elsewhere in the world, you may be able to have the date on which you applied for the earlier design accepted as the date on which you filed the later application. Priority dates may only be granted in relation to countries which have signed up to the Paris Convention or which are Members of the World Trade Organisation (WTO).

Using the Hague System

The Hague System for the International Registration of Industrial Designs allows you to simultaneously apply for a design in many different countries or territories, through a single application to the World Intellectual Property Organisation (WIPO). An application under the Hague System will cover the UK if the EU is selected.

****

Ch. 6

Copyright Protection

In this chapter, we will look at the protection of copyright. Contrary to other forms of protection, such as patents, trademarks and designs, the Intellectual property Office, other than offering advice, offers no system for protection of copyright. There is no registration system to provide protection. There is, however, lots of useful information on their website www.ipo.gov.uk

Definition of copyright

Copyright is the right to prevent others copying or reproducing an individuals or other's work. *Copyright protects the expression of an idea and not the idea itself.* Only when an idea is committed to paper can it be protected. Others can be directly or indirectly stopped from copying the whole or a substantial part of a copyright work. However, others cannot be stopped from borrowing an idea or producing something very similar.

Copyright is a right that arises automatically upon the creation of a work that qualifies for copyright protection. This means that there is no registration certificate to prove ownership. To claim ownership the author will have to produce original and preferably dated evidence of the creation of the work and proof of authorship. The author will also need to show that he is a qualifying person and that the work was produced in a convention country.

To be a qualifying person (s.154 of the Copyright Designs and Patents Act 1988) the author must have been, at the material time, a British Citizen, subject or protected person, a British Dependant territories citizen, a British national (overseas) or a British Overseas Citizen or must have been resident or domiciled in a convention country at the material time, which is when the work was first published. If the author dies before publication the material time is before his death. A convention country is a country that is signatory to the Universal Copyright Convention or the Berne Copyright Convention, which includes most countries in the world.

The works that can qualify for protection are defined in S.1 of the 1988 Act. These are:

a) Original literary, dramatic, musical and artistic works
b) Sound recordings, films, broadcasts and cable programmes
c) Typographical arrangements of published editions

Historical background

Copyright has its origins in the 16th century. The courts recognised a need for some form of protection for books. In 1556, a system of registration of books was established to offer protection for authors. If an author registered a book with the Stationers Company it gave him/her a perpetual right to reproduce the book and prevent reproduction by anyone else. For almost 200 years this form of protection only applied to books. In 1734 this extended to engravings (Engravings Copyright Act) A number of Acts were passed over the next 150 years extending copyright protection to musical, dramatic and artistic works. In 1875, a Royal Commission was set up to look at the position and recommended a clear approach be adopted to copyright protection, codified into one single Act. This happened after Great Britain signed the Berne Copyright Convention in 1885.

The Berne Convention provided for international protection of copyright for the work of all nationals of all countries signing the convention. It also required each member country to extend minimum standards of protection to nationals of all other member countries.

The United Kingdom implemented the 1911 Copyright Act to put into place minimum standards and also draw together previous legislation. The next Act, prompted by changes in the Berne Convention led to the 1956 Copyright Act. This Act reflected changes, amongst other things, in the field of technology. In 1973, the Whitford Committee was appointed to review the state of copyright law. The Committee reported in 1977 suggesting numerous changes to the law, resulting in a Green paper in 1981, 'Reform of the law relating to Copyright, Designs and Performers Protection' and subsequently the White Paper 'Intellectual Property and Innovation' which led to the 1988 Copyright Designs and Patents Act, which was a consolidating Act.

Since the Act came into force in August 1989, there have been a number of amending regulations dealing with implementation of EC Directives on rights to reproduce copyright software as is necessary for lawful use, protection of semiconductor chip typography rights and harmonisation of copyright duration. There are further legislative moves afoot to update copyright law to deal with the growth of new technology.

Copyright – Subsistence of copyright

As shown above, copyright is a property right that subsists in certain works. It is a statutory right giving the copyright owner certain exclusive rights in relation to his or her work.

In the 1988 Copyright Designs and Patents act there are nine categories of copyright works:

'Authorial' 'Primary' or 'LDMA' works

Literary works
1) Dramatic works
2) Musical works
3) Artistic works

'Entrepreneurial' 'Secondary' or 'Derivative' works
Sound recordings
4) Films
5) Broadcasts
6) Cable programmes
7) Typographical arrangements of published editions (the typography right)

Copyright comes into existence, or subsists automatically where a qualifying person creates a work that is original and tangible (or fixed).

Qualification
Copyright will not subsist in a work unless:

a) It has been created by a qualifying person
b) It was first published in a qualifying country
c) In the case of literary, dramatic and musical works, the work must be fixed, that is reduced to a material form in writing or otherwise.

Copyright works
The CDPA 1988 defines a literary work as being 'any work written, spoken or sung, other than a dramatic or musical work'. A novel or poem could equally fall into this category. Additionally, the concept of literary works extends to tables (e.g. a rail timetable) compilations such as directories and computer programmes. Databases are also regarded as literary works. In essence, any work that can be expressed in print, irrespective of quality, will be a literary work.

Dramatic works
The CDPA 1988 defines 'dramatic works' as including works of dance or mime.

Musical works

A musical work is a work consisting solely of musical notes, any words or actions intended to be sung, spoken or recorded with the notes are excluded. Therefore, a melody is a musical works with the lyrics being literary.

Artistic works

A wide-ranging definition of artistic works is provided by the CDPA 1988 s.4. Works of architecture are included but focus is usually placed on the remaining artistic works. These fall into two categories:

a) Works protected irrespective of their artistic merit:

    i) Graphic works, i.e. paintings, drawings, diagrams, maps, charts, plans, engravings, etchings, lithographs, woodcuts or similar works

    ii) Photographs

    iii) Sculptures. The protection of functional objects, such as a cast is problematic. In one notable case in New Zealand Wham-O manufacturing Co v Lincoln Industries Ltd (1985) a wooden model of a Frisbee was held to be a sculpture. The modern UK position is almost certainly more restrictive, as objects will not now be protected as sculptures where they are not made for the purpose of sculpture.

    iv) Collages. Collages are artistic or functional visual arrangements produced by affixing two or more items together. Intrinsically ephemeral arrangements (for example the composition of a photograph) are not collages.

b) Artistic works required to be of a certain quality (CDPA 1988 s.4 (1) c i.e. works of artistic craftsmanship. Few works can meet the standard of artistic craftsmanship, as they must be both of artistic quality and the result of craftsmanship. These principles were further developed into a two-part test for artistic craftsmanship in Merlet v Mothercare (1986). First, did the creation of the work involve craftsmanship in the sense that skill and pride was invested in its manufacturer? Second, does the work have aesthetic appeal and did an artist create it?

Sound recordings

A sound recording is a reproducible recording of either:

1) Sounds where there is no underlying copyright work (e.g. birdsong)

2) A recording of the whole or any part of a literary, dramatic or musical work.

The format of recording is of no relevance.

Film

The CDPA 1988 s.5B (1) provides that a film is a reproducible recording of a moving image on any medium. It is the recording itself that is protected, rather than the subject matter that has been recorded, but it should be borne in mind that a film might also be protected as a dramatic work. Film soundtracks are taken to be part of the film itself.

Broadcasts

Copyright subsists in sounds and visual images that are broadcast CDPA 1988 s.6 (1), a broadcast being defined as a transmission by wireless telegraphy of visual images, sounds or other information. The definition of 'broadcast' therefore encompasses radio and television broadcasts and both terrestrial and satellite broadcasting.

Cable programmes

The transmission of an item that forms part of a cable programme will create separate works that are capable of protection as cable programmes CDPA 1988 s.7. A cable programme service is defined as a service consisting wholly or mainly in sending visual images, sounds or other information via a telecommunications system which may utilise wires or microwave transmission. Items sent via wireless telegraphy are specifically excluded as they are already protected as broadcasts. This means that as well as subscription channels a website on the internet may be a cable programme service.

The typography right

The CDPA 1988 s.8 affords protection to the typography, that is the layout, of published editions of literary, dramatic and musical works.

Copyright works the ideas/expression dichotomy

There is no copyright in ideas. Copyright subsists in the tangible expression of ideas and not the ideas themselves. In America this is referred to as the ideas/expression dichotomy. This principle can be helpful but should not be taken too literally, as whilst it is clear that mere ideas cannot be protected by copyright the following points should be noted:

What might be termed 'highly developed ideas', for example an early draft of a textbook, would be protected by copyright, as are preparatory design materials for computer programmes.

1) Copyright cannot be circumvented by selectively altering the expression of a copyright work in the process of reproducing it.

Originality

The CDPA 1988 s.1 requires that literary, dramatic, musical and artistic works be 'original'. The originality requirements only apply to LDMA works, there is no such requirement for secondary copyright works, although it is clear that no copyright will subsist in secondary copyright works that merely reproduce secondary works.

LDMA works must be original in the sense that they originate with the author. Expending skill and judgement in creating an LDMA work usually suffices to deem the work original. Mere copying cannot confer originality. Alternatively, the mere expenditure of effort or labour (the so-called 'sweat of the brow' test for originality) has sometimes been said to be sufficient to confer originality. But in practice some minimum element of originality is required. For example, in Crump v Smythson (1944) it was held that the generic nature of commonplace diary material left no room for judgement in selection and arrangement therefore the resultant works were not original. Originality has also been held to be more than 'competent draftsmanship' (Interlego v Tyon 1988). Commonly databases and computer programmes were the subject matter of sweat of the brow concerns.

Higher standards of originality: computer programs and databases

As a result of two European Directives, The Directive on the Legal Protection of Databases (Directive 96/9/EC) and the Computer Directive (Directive 91/250/EEC) both computer programmes and databases must be original in the sense that they are the author's own intellectual creation. This is a higher standard or originality than that of 'skills, labour and judgement'.

Some databases may not meet the standard of originality to be afforded copyright protection. In this case the database can be protected by virtue of the *sui generis* database right.

The Database Directive which was incorporated into UK law by Part 111 of the Copyright and Rights in Databases Regulations 1997 grant a property right in a database whether or not it qualifies for a copyright work. The definition of database includes:

' a collection of independent works, data or other materials arranged in a systematic or methodical way and individually accessible by electronic or other means'.

A database can also be recognised as a literary work and thus afforded copyright protection. For this the database must be original and the contents and arrangements of the database must be a result of the author's own intellectual creation. In any case, all databases are protected by the new database rights irrespective of whether they qualify for copyright protection or not. To qualify for database rights the data must have been assembled through substantial investment in obtaining, verifying and presenting the contents. The duration of the database rights is for 15 years from 1st January of the year following completion of its making, or the first making public of the database within the 15 year period from its making.

Fixation and Tangibility

As we have seen, copyright does not subsist in literary, dramatic or musical works until they are recorded in writing or otherwise. This pragmatic requirement is known as 'fixation'. Usually, such works will be fixed by the author, but fixation by a third party (with or without the authors permission) is also possible.

Other copyright works are not subject to the fixation requirement. This is usually unproblematic as films, sound recordings, broadcasts, cable programmes and typography are inherently tangible works.

Ownership of copyright and the employee

The rule is that the first owner of copyright in a work is the person who created the work, i.e. the author. A major exception to this rule is CDPA 1988 s.11 (2). Which provides that where a person creates an LDMA work in the course of employment the employer is the first owner of any copyright in the work subject to any agreement to the contrary. There are special provisions for Crown use, Parliamentary copyright and copyright for certain international organisations (CDPA 1988 s.11 (3).

Authorship, ownership and moral rights

The author is the person who creates the work. Identifying the author is usually a straightforward task. The following is the standard authorship position:

- Literary work. The writer
- Dramatic work. The writer
- Musical work. The composer
- Artistic work. The artist
- Computer generated LDMA works. The person operating the computer.
- Sound recordings. The producer.
- Films. The producer and principal director.
- Broadcasts. The broadcaster.
- Cable programmes. The cable program service provider.
- Typography right. The publisher.
- Any work where the identity of the author is unknown. A work of unknown authorship.

Joint authorship

Where more than one person is involved in the creation of a work, careful consideration is needed in determining individual contributions. A person who suggests a subject to a poet is not the author of the poem. Merely supplying ideas is insufficient for joint authorship; an integral role in the expression of ideas is required. Joint authorship arises where the efforts of the two authors is indistinguishable.

In the case of infringement the claimant may wish to apply for an interlocutory injunction, because the continued reproduction of infringing articles pending a full hearing could put the copyright owner out of business or be prejudicial in some other way.

An exclusive copyright licensee will have the same rights of the copyright owner in respect of an infringement committed after a licensee has been granted. With the exception of an interlocutory injunction, which the exclusive licensee must bring alone all other actions by a licensee must be brought in conjunction with the copyright owner.

In proceedings relating to copyright infringement, there are a number of presumptions laid down by the 1988 Act (in ss27(4) 104, 105 and 106) that allow certain issues to be assumed and that shift the burden of proof to the other party.

Copyright infringement

Section 16(1) and (2) Copyright, Designs and Patents Act 1988 states:

"The owner of the copyright in a work has the exclusive right to copy, issue copies of the work, rent, lend, perform, show, play or communicate the work to the public or do any of the above in relation to an adaptation. Copyright in a work is infringed by a person who without the licence of the copyright owner does, or authorises another to do, any of the acts restricted by copyright".

However, the 1988 Act has been significantly amended by a series of Regulations introduced in June 2014 These changes affect infringement of copyright.

Amendments to Copyright law-New Regulations from 2014

The following Regulations have all been brought into force 1st June 2014 and have served to amend the Copyright, Designs and Patents Act 1988.

The Public Administration, Disability, and Research, Education, Libraries and Archives statutory instruments were approved by parliament on 14 May and came into force on 1 June 2014. These instruments will amend relevant sections of the Copyright, Designs and Patents Act 1988.

- The Copyright and Rights in Performances (Disability) Regulations 2014
- The Copyright and Rights in Performances (Personal Copies for Private Use) Regulations 2014
- The Copyright and Rights in Performances (Research, Education, Libraries and Archives) Regulations 2014
- The Copyright (Public Administrations) Regulations 2014
- The Copyright and Rights in Performances (Quotation and Parody) Regulations 2014

Exceptions to copyright

Copyright protects literary, dramatic, musical and artistic works as well as films, sound recordings, book layouts, and broadcasts. If you want to copy or use a copyright work then you usually have to get permission from the copyright owner, but there are a few exceptions where you can copy or use part or all of a copyright work without permission. Where a work contains a performance, the performance will also have rights over how the work is used. The exceptions to copyright also apply to these related rights.

The law on these exceptions has changed in a number of small but important ways, to make the UK copyright system better suited to the digital age. These changes affect how you can use content like books, music, films and photographs.

Personal copies for private use
What has changed?

Copyright law has been changed to allow you to make personal copies of media (ebooks, digital music or video files etc) you have bought, for private purposes such as format shifting or backup. Before this change to the law, it was not legal to copy music that you bought on a CD onto your MP3 player. The changes, which apply from 1 October, update copyright law to make this legal, as long as you own what you are copying, e.g. a music album, and the copy you make is for your own private use.

You are also able to copy a book or film you have bought for one of your devices onto another of your devices, without infringing copyright. However, you should note that media, such as DVDs and e-books, can still be protected by technology which physically prevents copying and circumvention of such technology remains illegal. It is still illegal to make copies for friends or family, or to make a copy of something you do not own or have acquired illegally, without the copyright owner's permission. So you cannot make copies of CDs for your friends, copy CDs borrowed from friends, or copy videos illegally downloaded from file-sharing websites.

The law allows you to make personal copies to any device that you own, or a personal online storage medium, such as a private cloud. However, it is illegal to give other people access to the copies you have made, including, for example, by allowing a friend to access your personal cloud storage.

The exception applies to any copies you have bought, other than computer programs. So, for example, it allows you to format shift an ebook you have bought from one device to another for your own private use. However, you should note that media, such as DVDs and e-books, can still be protected by technology which physically prevents copying and circumvention of such technology remains illegal.

Quotation
What has changed?

Previously, it was an infringement of copyright to take a quotation from one work and use it in another without permission from the copyright owner, unless it was done for the purpose of criticism, review or news reporting.

Copyright law allows quotations to be used more widely without infringing copyright, as long as the use is fair (in law, the use must be a "fair dealing", see the box below) and there is a sufficient acknowledgement - which generally means the title and the author's name should be indicated. It is ultimately for the courts to determine whether use of a quotation is fair dealing, which will depend on the facts of any specific case. However, the use of a title and short extract from a book in an academic article discussing the book is likely to be permitted, whereas the copying of a long extract from a book, without it being justified by the context, is unlikely to be permitted. You may benefit from this law if you are an author, academic, or even just a casual blogger.

Caricature, parody or pastiche
What has changed?

Previously, anyone wishing to use other people's copyright material for the purposes of caricature, parody or pastiche (such as a parody song or video), required the permission of the rights holder. Copyright law now allows limited uses of copyright material for the purposes of caricature, parody or pastiche, without having to obtain the permission of the rights holder. It is important to ensure you understand the limits if you plan to use other people's material for caricature, parody or pastiche. Only minor uses are permitted and a use must be considered fair and reasonable (in law, the use must be a "fair dealing", see the box below), otherwise you must seek permission from the rights holder.

This exception to copyright has no impact on the law of libel or slander, so you may still be sued if a parody work is defamatory. It also does not affect an author's moral right to object to "derogatory treatment" of their work (as defined in copyright law).

What is fair dealing?

'Fair dealing' is a legal term used to establish whether a use of copyright material is lawful or whether it infringes copyright. There is no statutory definition of fair dealing - it will always be a matter of fact, degree and impression in each case. The question to be asked is: how would a fair-minded and honest person have dealt with the work?

Factors that have been identified by the courts as relevant in determining whether a particular dealing with a work is fair, include:

- Does using the work affect the market for the original work? If a use of a work acts as a substitute for it, causing the owner to lose revenue, then it is not likely to be fair.
- Is the amount of the work taken reasonable and appropriate? Was it necessary to use the amount taken? Usually only part of a work may be used.

The relative importance of any one factor will vary according to the case in hand and the type of dealing in question.

Education and teaching

Changes have been made to copyright law in order to help teachers to deliver modern multi-media teaching without risk of copyright infringement.

The exceptions relating specifically to educational establishments have widened, allowing more extensive use of materials in conjunction with educational licensing schemes.

Another change permits minor acts of copying for teaching purposes, as long as the use is considered fair and reasonable. So, teachers will be able to do things like displaying webpages or quotes on interactive whiteboards, without having to seek additional permission. Education and Teaching

What has changed?

Many schools, colleges and universities copy media which is protected by copyright - for instance photocopying extracts from books for class handouts or recording television programmes to show to a class.

In order to do this, educational establishments hold educational copying licences. So if a school wants to record television broadcasts, it needs a licence from the Educational Recording Agency. If it wants to photocopy extracts from books, it needs a licence from the Copyright Licensing Agency. Most educational establishments already hold these licences.

These licensing schemes are underpinned by copyright exceptions which mean that, where a particular work is not covered by a licence, an educational establishment is still able to copy it. This means that teachers do not have to check the terms of each item they want to copy before they copy it.

The changes to the law apply these exceptions to a wider range of copyright works which were previously not covered - such as artistic works (including photographs), films and sound recordings. They also permit sharing of copies over secure distance learning networks. In order to carry out these activities, schools, colleges and universities simply need to make sure they hold the relevant licences.

The previous law also allowed limited copying of literary, dramatic, musical or artistic works for the purposes of teaching, provided it was not by means of a reprographic process. This meant copying by hand was permitted, but the use of laptops and interactive whiteboards was not.

The previous law has been replaced with a general "fair dealing" exception, allowing copying of works in any medium as long as the following conditions apply:

1. the work must be used solely to illustrate a point;
2. the use of the work must not be for commercial purposes;
3. the use must be fair dealing; and
4. it must be accompanied by a sufficient acknowledgement.

This means minor uses, such as displaying a few lines of poetry on an interactive whiteboard, are permitted, but uses which would undermine sales of teaching materials still need a licence.

The new law does not remove the need for educational establishments to hold licences for use that does not fall under the "fair dealing" exception, for instance, photocopying material to distribute to students. Schools, colleges and universities still have to pay for third party teaching materials which are available under licence.

Teaching use which is fair dealing, illustrative and non-commercial is permitted by the exception and uses for exam purposes may fall within these criteria. You may also be able to rely on the new quotation exception, for example where you wish to reproduce a piece of text for analysis in an English exam. This would not extend to the making of a reprographic copy of a musical work for use by an examination candidate when performing the work.

Research and Private study
What has changed?

The change in the law means that for the first time, researchers and students who need to copy parts of sound recordings, films or broadcasts for non-commercial research or private study are allowed to do so. Libraries and archives are also able to make copies of artistic works for researchers and students. Education institutions, libraries, archives and museums are able to offer access to copyright works on their premises by electronic means at dedicated terminals.

Researchers and students were previously allowed to copy limited extracts of literary, dramatic, musical and artistic works for non-commercial research and private study. They are now able to copy a limited amount of a sound recording, film or broadcast. This amount is restricted by fair dealing, which rules out unfair or unreasonable uses such as copying a whole film for "research" instead of buying the DVD. Any use made of the work must be accompanied by a sufficient acknowledgement.

Quotation
What has changed?

Before the law changed, minor uses of quotations from copyright works could be prevented by copyright owners, unless they fell within fair dealing exceptions for criticism, review or news reporting.

The law has been amended to give people greater freedom to quote the works of others for other purposes, as long as this is reasonable and fair ("fair dealing").

Caricature, parody or pastiche
What has changed?

The ability to re-edit copyright works in new and experimental ways is seen as an important learning and teaching exercise for creative skills.

Many works of caricature, parody or pastiche, involve some level of copying from another work. The law has changed to allow limited uses of other people's copyright material for the purposes of caricature, parody or pastiche, without first asking for permission.

Changes affecting disabled people

The Government has made it easier for disabled people to access materials that are protected by copyright. Previously there were exceptions to copyright law that allowed visually-impaired people, and organisations acting on their behalf, to make accessible versions (e.g. Braille versions) of certain types of material, such as books. Also, the law allowed certain designated organisations to produce sub-titled copies of broadcasts for people who are deaf or hard of hearing.

However, these exceptions did not apply to other types of impairment, such as dyslexia, and did not apply to all types of copyright work. The law has changed so that anyone who has an impairment that prevents them accessing copyright works will now be able to benefit from the exception, not just visually-impaired people. The law also allows individuals, educational establishments and not-for-profit organisations to reproduce all types of copyright-protected content in accessible formats.

These changes will give disabled people greater access to creative content, as individuals, educational establishments and charities are allowed to reproduce material without purchasing a licence as long as a copy is not already commercially available in an accessible format.

Also, the law has been simplified so that organisations wishing to produce sub-titled copies of broadcasts on behalf of deaf and other disabled people will be able to do so without going through a bureaucratic designation process.

What's allowed?

The change in the law makes the rules on what can be copied into an accessible format much broader so that a whole range of content, including films and broadcasts, can be reproduced in an accessible format for disabled people. The law allows acts such as:

- Making Braille, audio or large-print copies of books, newspapers or magazines for visually-impaired people
- Adding audio-description to films or broadcasts for visually-impaired people
- Making sub-titled films or broadcasts for deaf or hard of hearing people
- Making accessible copies of books, newspapers or magazines for dyslexic people

However, it is only legal to reproduce material if suitable accessible copies are not commercially available.

Organisations that make and supply accessible-format copies for disabled people have a duty to keep records of the copies they make and provide them to the copyright owner of the material.

Research
What's changed?

The law already permitted limited copying of some types of copyright material, such as books, for non-commercial research or genuine private study. The law has now changed so that all types of copyright works are covered.

Text and data mining technologies help researchers process large amounts of data.

Copyright law has altered to help ensure that if a researcher is carrying out non-commercial research they will not infringe copyright by copying material for a text and data mining analysis.

The conditions under which this exception applies are set out in this guidance.

Copying works for research and private study

Copyright law recognises that researchers and students may legitimately need to copy limited extracts of copyright works for the purpose of their studies. Therefore, the law already allowed researchers and students to copy limited extracts of some types of copyright works (books, plays and musical scores, picture and photos, literary, dramatic musical and artistic works) as long as they are carrying out non-commercial research or private study. Librarians are permitted to assist researchers and students by providing limited copies of these types of copyright works.

What has changed?

Copyright law has now changed so that all types of published copyright works are covered by the exception allowing limited copying for the purpose of research. This means that researchers and students (or the librarians and archivists who are assisting them) who need to copy limited parts of sound recordings, films or broadcasts for non-commercial research or private study are allowed to do so. The provisions about only copying a part of a work and sufficiently acknowledging the original work still apply.

Additionally, educational institutions, libraries, archives and museums are now permitted to offer access to copyright works on their premises at dedicated electronic terminals for research and private study. More information can be found in the leaflet "Exceptions to copyright: Libraries, archives and museums".

Text and Data Mining for non-commercial research

Text and data mining is the use of automated analytical techniques to analyse text and data for patterns, trends and other useful information. Text and data mining usually requires copying of the work to be analysed. Before the law was changed, researchers using text and data mining in their research risked infringing copyright unless they had specific permission from the copyright owner.

What's changed?

The new copyright exception allows researchers to make copies of any copyright material for the purpose of computational analysis if they already have the right to read the work (that is, work that they have "lawful access" to). They will be able to do this without having to obtain permission to make copies from the rights holder. This exception only permits the making of copies for the purpose of text and data mining for non-commercial research. Researchers will still have to buy subscriptions to access material; this could be from many sources including academic publishers.

Publishers and content providers are able to apply reasonable measures to maintain their network security or stability so long as these measures do not prevent or unreasonable restrict a researcher's ability to make the copies they need to make for their text and data mining. Contract terms that stop researchers making copies of works to which they have lawful access in order to carry out a text and data mining analysis will be unenforceable.

Quotation
What has changed?

Before the law changed, minor uses of quotations from copyright works could be prevented by copyright owners, unless they fell within fair dealing exceptions for criticism, review or news reporting.

The law has been amended to give people greater freedom to quote the works of others for other purposes, as long as this is reasonable and fair ("fair dealing).

Copyright material held by public bodies
What's changed?

Some material held by public bodies will have been submitted by third parties, such as by members of the public, businesses or researchers. For example this could be material submitted to a public body, such as local authority, as part of a duty to capture information required for public register.

Where a public body holds third party material - that is, material in which someone other than the public body owns the copyright - the general position has been that an individual could either view that material in person, or a public body could copy and distribute it on an individual basis. The law did not generally allow such material to be published online without permission from the rights holders.

Public bodies and keepers of statutory registers are now allowed to proactively share copyright material online without seeking permission, as long as it is not commercially available. The same applies to material that is already available for public inspection through some statutory mechanism, such as local planning applications.

This change does not permit public bodies to publish material that is commercially available to buy or license (such as academic articles). In these circumstances any public body would still need to seek the permission of the rights holder.

Another change in this area applies to certain works that have been communicated to the Crown with the permission of the copyright owner and in the course of public business. This exception applies only to literary, dramatic, musical or artistic works that have not previously been published.

The change to the law will make it easier for the public to access information, saving both time and expense for public bodies and individuals.

Libraries, Museums and Archives
What has changed?
The law has changed to make it easier and cheaper for cultural institutions like libraries, archives, and museums to use, share and preserve their collections.

There are two significant changes which will affect libraries, archives and museums.

The first relates to making copies of works to preserve them for future generations.

The second allows greater freedom to copy works for those carrying out non-commercial research and private study.

Archiving and preservation of resources
What's changed?
Changes to copyright law for archiving and preservation will make it easier to preserve creative content held by libraries, archives and museums. These institutions are now allowed to preserve any type of copyright work that is held in their permanent collection (but not available for loan to the public) and cannot readily be replaced.

Why is it necessary to allow multiple copies of work to be made for preservation/archiving?
There is a risk that both the original and the copy of a work may degenerate or corrupt over time. Making a single copy may be insufficient to safeguard a work in the long term.

Research and private study
Copying works for research and private study
Librarians are permitted to assist researchers and students by providing limited copies of books, plays and musical scores, pictures and photos, literary, dramatic, musical and artistic works for non-commercial research and private study. The amount that can be copied is restricted to a reasonable proportion. This rules out unfair

or unreasonable uses such as copying a whole film for "research" instead of buying the DVD and generally means that only a part of a work can be copied. Use made of the work should be accompanied by sufficient acknowledgement (e.g. in a reference or bibliography).

What's changed?

The law has changed so that all types of published copyright works are now covered by the provisions in copyright law allowing limited copying for non-commercial research and private study. The same provisions about only copying a part of a work and sufficiently acknowledging the author still apply.

Educational institutions, libraries, archives and museums are now permitted to offer access to copyright works on their premises at dedicated electronic terminals for research and private study.

How does a librarian ensure that the person is genuinely doing non-commercial research or private study?

A librarian who is supplying a copy of a work will wish to ask a researcher to declare that they are doing non-commercial research, this can now be done electronically (for example, an electronic copyright declaration form could be signed using a typed signature or check-box).

What types of libraries are able to provide copies of work?

Publicly accessible libraries and archives, such as those run by universities, schools, local council, government departments and NHS institutions can provide copies of copyright work for non-commercial research and private study, as set out in this guidance. The provisions do not extend to private libraries or archives, such as law firms that run on a commercial basis.

Libraries, archives and museums

How much of a work is a librarian allowed to copy for a student or researcher?

The amount you are able to copy of a published work is limited to a reasonable proportion, and a copyright declaration must be provided. This generally means that only a limited part that is necessary for the research project may be copied.

However, archivists may supply a single copy of a whole or part of a work provided that the work had not been published or communicated before it was deposited in the library or archive, or the owner of that work has not prohibited the copying of that work. As usual, a copyright declaration must be provided.

Defences to Copyright infringement

There are a number of defences to infringement:
a) Challenge the existence of copyright or the claimant's ownership of copyright.
b) Deny the infringement.
c) Claim to have been entitled, because of permission granted to do the act in question or argue that it is within one of the statutory fair dealing exemptions or by claiming public interest or EC competition rights.

A claim of ignorance of the law will not work as a defence. Ignorance of subsistence of copyright will, however, have a bearing on any damages awarded. In the case of secondary infringement an element of knowledge is required for the infringement to be actionable in the first place. The infringement only occurs if the person knows that what he or she is dealing with is an infringing copyright work.

If the claimant does own copyright in the work that is allegedly infringed, and facts can be proved, the only defences remaining are:

1) that the defendant had permission from the copyright owner to make a copy.

Provided that the defendant in an infringement action can prove that permission was granted, either in writing, orally or, in certain cases, implied, then the claim of infringement will fail.

2) That the act was one of the permitted acts under the 1988 Act, as amended

The 1988 Act, as amended, contains statutory permissions, or exceptions, to the exclusive rights of the copyright owner. Many of these have come from the results of case decisions over the years that have acknowledged the need for fair exceptions. These permitted acts are categorised in the Act and comprise:

- Research and private study
- Criticism review and news reporting
- Incidental inclusion of copyright material
- Things done for instruction or examination
- Anthologies for educational use
- Playing, showing or performing in an educational establishment
- Recordings by educational establishments
- Reprographic copying by educational establishments
- Libraries and archives
- Public administration
- Lawful users of computer programs and databases
- Designs

- Typefaces
- Works in electronic form

All of the above are categorised in the Act and each case concerning these categories will be on its own merit.

3) That the exercise of the copyright owner's rights to prevent copying would amount to an anti-competitive practice under EC competition law.
4) That the exercise of the copyright owner's rights is against the public interest.

Each one of the above must be proven and each case will be judged on its own merit.

****

6

Intellectual Property and Computer Software

Computer software and computer designs have pushed the boundaries of intellectual property law.

All forms of copyright work can exist in digital form and the Copyright, Designs and Patents Act 1988 acknowledges this, for example, by providing that copying includes copying in electronic form and also extends to transient and temporary copies. protection is also given to technological measures to prevent unauthorised acts in relation to computer programmes (in this case called copy-protection) and other forms of works, including databases, have suitable provisions for this. Of particular interest is the protection of computer programs and databases, which may also be protected by a database right.

Computer programs

Computer programs and preparatory design material for computer programs are a form of literary work. Section 3(1) Copyright, Designs and Patents Act 1988 states:

...'literary work' means any work, other than a dramatic or musical work, which is written, spoken or sung, and accordingly includes -

(i) a table or compilation (other than a database);(ii) a computer program;
(iii) preparatory design material for a computer program; and
(iv) a database

For the purposes of copyright a computer programme is original in the sense that it is the author's own intellectual creation. the Act simply requires that a computer program is original, unlike the Directive which uses the above formula.

The basic rules on the restricted acts and infringement apply to computer programs as with other forms of literary work. However, there are some special rules as to what constitutes an adaptation of a compute program and there are some specific permitted acts that apply to computer programs (back-up copies, decompilation, observing, studying and testing and other acts permitted to lawful users). Rules as to ownership are the same as other original works but the nature of the software industry means that many computer programs and other items of software are created by consultants and self-employed persons.

There are no significant issues with respect of duplicate copying of computer programs but non-textual copying can cause problems, such as where an alleged infringer has written a new computer program to emulate the operational and functional aspects of an existing program, sometimes using a completely different computer programming language.

On case that highlights this is Nova Productions Ltd v Mazooma Games Ltd (2007) where the defendant created a computer game which had some similar features to the claimant's game such as the 'power cue'. Both games were based on the game of pool with coloured balls and a green baize table with pockets. The main finding here was that there was no infringement. Although there were similarities, there was no allegation of copying the claimant's program source code. merely emulating an existing program without more does not infringe. the claim was not sufficiently specific to cover copying of the detailed architecture of the claimant's program and merely amounted to a claim to copying general ideas. Jacob LJ said:

'If protection for such general ideas as are relied on here were conferred by the law, copyright would become an instrument of oppression rather than the incentive for creation which it is intended to be. protection would have moved to cover works merely inspired by others, to ideas themselves'.

Databases

It should be noted that although 'computer program' is not defined in the act, there is a definition of 'database' that applies equally to copyright databases and databases protected by the database right even though the two rights are different.

The main definition of a database is that a database is a collection of independent works, data or other materials which are arranged in a systematic or methodical way and are individually accessible by electronic or other means.

Databases can be electronic or other wise. For example, a card index arranged alphabetically would fall within the definition. The contents themselves need not be works of copyright (data or other materials) but if the contents are protected by copyright or other rights, such rights are not prejudiced by the protection of a database as a database.

For the database right to subsist, there must be substantial investment (in terms of human, technical or financial resources) in the obtaining, verification or presentation of the contents of the database. The repeated and systematic extraction and re-utilisation of insubstantial parts of the contents of a database may infringe the database right where, cumulatively. these acts amount to an extraction and/or re-utilisation of a substantial part of the contents of a database measured qualitatively and/ or quantitatively.

Patents and computer software

Certain things are excluded from the meaning of 'invention' for the purposes of patent law. the list of things is not exhaustive, nor is there any deducible common logic between the things excluded. Computer programs, as such, are excluded as are business methods and mental acts. Where there is a 'technical effect' this may overcome the exception but where a computer program only performs other excluded matter, this will not count as a valid technical effect, for example, in the case of a business method implemented by a computer program.

The main statute covering this area is Article 52 of the European patent Convention. This states:

(1) European patents shall be granted for any inventions, in all fields of technology, provided that they are new, involve an inventive step and are susceptible of industrial application.
(2) The following in particular shall not be regarded as inventions within the meaning of paragraph 1:

(a) discoveries, scientific theories and mathematical methods;
(b) aesthetic creations;
(c) schemes, rules and methods for performing mental acts, playing games or doing business, and programs for computers;
(d) presentation of information.

(3) paragraph 2 shall exclude the patentability of the subject matter or activities referred to therein only to the extent to which a European patent application or European patent relates to such subject matter or activities as such.

Design law and computer generated images

Before the harmonisation of registered design law in Europe and the introduction of the Community design, it was almost impossible to register computer graphics as designs in the UK. The definition of 'design' and 'product' under the harmonising directive and the Community Design Regulation changed this. The key statute here is Article 3 (a) and (b) of the Community Design Regulation (Article (1(a) and (b) of the Directive on the legal protection of designs)

(a) 'design' means the appearance of the whole or a part of a product resulting from the features of, in particular, the lines, contours, colours, shape, texture and/or materials of the product itself and/or its ornamentation;

(b) 'product' means any industrial or handicraft item, including *inter alia* parts intended to be assembled into a complex product, packaging, get up, graphic symbols and typographic typefaces, but excluding computer programs.

Although computer programs are expressly excluded, this does not extend to symbols and images generated by computer programs. Also, fonts, being typographical typefaces, are capable of protection.

****

7

Other protection

Intellectual Property (IP) covers a wide range of subject areas and you may find that you can protect your idea by another right.

Companies House
Companies House deals with the registration and provision of company information.

Registering company names
Companies House is responsible for company registration in Great Britain.

Company law is different from trade mark law. You cannot stop someone using a trade mark, which is the same or similar to yours, just by registering your name with Companies House.

The IPO cannot guarantee that the name of a company accepted for registration at Companies House is acceptable by the IPO as a registered trade mark.

The company name may not qualify as a trade mark because, for example:
- It is not considered distinctive,
- It is a descriptive word or term
- It may indicate geographical origin,
- It may already be registered in someone else's name

The following examples of company names would not be accepted as trade marks:

- Trutworthy plumbers
- Cheap insurance

In the same way, a trade mark, which is a word, might not be accepted for registration at Companies House.

Company Names Tribunal

The Tribunal adjudicates in disputes about opportunistic company name registrations.

Conditional access technology

For encrypted broadcasts and transmissions, you may need Conditional access technology.

Conditional access technology generally refers to technical measures, such as smart cards or other decoders, which allow users to view or listen to encrypted broadcasts.

Some broadcasts and other transmissions are in an encrypted form so that they can only be seen by a person who has the right decoding equipment, a system usually used when broadcasters wish to charge recipients of the transmission.

On payment of the appropriate fee a person is given or is entitled to use a decoder and view the transmission.

In the same way that people make illegal copies of copyright works, they may make unauthorised smart cards or other decoding equipment with the intention of selling them in competition with the legitimate decoders, and so depriving the broadcaster or cable operator of the payments that would normally be paid for reception of the transmissions.

The law therefore sets out in what circumstances it is illegal to make and sell or otherwise deal in unauthorised decoders: there may be criminal as well as civil penalties. If you use an illegal decoder to receive broadcasts you're not entitled to, you may be committing an offence.

The Telecommunications UK Fraud Forum (TUFF) represents some makers of encrypted transmissions who are concerned about illegal decoders in the United Kingdom.

Copy protection devices

For copyright material issued to the public in an electronic form, you may decide to use technical measures so that it is not possible to make a copy of your material, that is, it is copy-protected.

It is also possible for you to use other technological measures to prevent other types of illegal uses of copyright material.

Where you have sold copies that are protected by technical measures, you may have the right to take action against a person who gets round or who makes, sells or otherwise deals in devices or means specifically designed or adapted to get round, the technical measures.

The right to take action is equivalent to the rights you have when suing for infringement of your copyright in the civil courts. Criminal offences may also apply to those who deal in the means to get round technical measures

Confidentiality agreements (CDAs)

It is important that you do not make your invention public before you apply to patent it, because this may mean that you cannot patent it, or it may make your patent invalid.

However, that does not mean that you must never discuss your invention with anyone else. For example, you can discuss it with qualified (registered) lawyers, solicitors and patent attorneys because anything you say to or show them is legally privileged. This means it is in confidence and they will not tell anyone else.

Alternatively, you may need to discuss your invention with someone else before you apply for a patent – such as a patent adviser or consultant, or an inventor-support organisation. If so, a Non-Disclosure Agreement (NDA) can help. NDAs are also known as confidentiality agreements and confidentiality-disclosure agreements (CDA).

No single NDA will work in every situation. This means that you must think carefully about what to include in your NDA. You may want to consult a qualified lawyer or patent attorney if you are thinking about discussing your invention with someone else and are considering using a non-disclosure agreement.

Plant breeders rights

If you have created a new variety of plant or seed, you may be able to protect it at The Plant Variety Rights Office and Seeds Division in the Department for Environment Food and Rural Affairs (DEFRA).

Publication right

Publication right gives rights broadly equivalent to copyright, to a person who publishes for the first time a literary, dramatic, musical or artistic work or a film in which copyright has expired. However, there is one major difference, publication right only lasts for 25 years from the year of publication of the previously unpublished material. It is important to note that the owner of publication right is the person who first publishes the unpublished material in which copyright has expired which will not necessary be the original owner of the copyright in the work.

This right should not be confused with the protection afforded published editions

Protection abroad

If you want to protect your IP abroad you will generally need to apply for protection in the countries which you want your IP to have effect. Except particularly in the case of Copyright and other limited circumstances, your UK IP rights do not give you automatic protection abroad .

Trade secrets

You should consider keeping something as a trade secret if:

- it is not appropriate for intellectual property (IP) protection
- you want to keep it secret or
- you want protection to extend beyond the term of a patent

If it would be difficult to copy the process, construction or formulation from your product itself, a trade secret may give you the protection you need.

However, a trade secret does not stop anyone from inventing the same process or product independently, and can be difficult to keep. The law of confidentiality protects trade secrets. To keep trade secrets protected, you must establish that the information is confidential, and ensure that anyone you tell about it signs a Non-Disclosure agreement (NDA). If they then tell anyone about it, this is a breach of confidence and you can take legal action against them.

****

Useful Addresses and Websites

Chartered Society of Designers
www.csd.org.uk

Chartered Institute of Patent Attorneys
www.cipa.org.uk

Companies House
www.companieshouse.gov.uk

England and Wales
Companies House
Crown Way
Cardiff CF14 3UZ

Espacenet (search existing paetnts)
www.epo.org/searching/free/espacenet.html

European Patent Office
www.epo.org

FICPI-UK
The national United Kingdom association of the International Federation of Intellectual Property Attorneys.
www.ficpi.org.uk

Intellectual Property Office
Concept House
Cardiff Road
Newport
South Wales
NP10 8QQ
Telephone +44 (0)1633 814184

Fax +44 (0)1633 817777
www.ipo.gov.uk

Institute of Patentees and Inventors
www.invent.org.uk

Institute of Trade Mark Attorneys
www.itma.org.uk

Nominet (Domain name searches)
www.nominet.org.uk

World Intellectual Property Organization
www.wipo.int

****

Index

****

Appendix 1

Standard forms

Application for Patent (PF1)

Application for Trade Mark (TM3)

Application for Registered Design (DF2A)

Intellectual Property Office

Patents Form 1
Patents Act 1977 *(Rule 12)*

Request for grant of a patent
(An explanatory leaflet on how to fill in this form is available from the office)

Concept House
Cardiff Road
Newport
South Wales
NP10 8QQ

Application number GB

1	Your reference: *(optional)*	

2	Full name, address and postcode of the applicant or of each applicant *(underline all surnames):* **The name(s) and address(es) provided here will be published as part of the application process (see warning note below)** Patents ADP number *(if you know it):*	

3	Title of the invention:	

4	Name of your agent *(if you have one):* "Address for service" to which all correspondence should be sent *(including postcode)*. This may be in the European Economic area or Channel Islands: (see warning note below) Patents ADP number *(if you know it):*	

	Country	Application number *(if you know it)*	Date of filing *(day / month / year)*
5 Priority declaration: Are you claiming priority from one or more earlier-filed patent applications? If so, please give details of the application(s):			

	Number of earlier UK application	Date of filing *(day / month / year)*
6 Divisionals etc: Is this application a divisional application, or being made following resolution of an entitlement dispute about an earlier application? If so, please give the application number and filing date of the earlier application:		

7	Inventorship: (Inventors must be individuals not companies)	(Please tick the appropriate boxes)	
	Are all the applicants named above also inventors?	YES ☐	NO ☐
	If yes, are there any other inventors?	YES ☐	NO ☐
8	Are you paying the application fee with this form?	YES ☐	NO ☐

9   Accompanying documents:
please enter the number of pages of
each item accompanying this form:

Continuation sheets of this form:

Description:

Claim(*s*):

Abstract:

Drawing(*s*):

If you are <u>not</u> filing a description, please   Country   Application number   Date of filing
give details of the previous application   *(day / month / year)*
you are going to rely upon:

---

10   If you are also filing any of the following,
state how many against each item.

Priority documents:

Statement of inventorship and right
to grant of a patent *(Patents Form 7)*:

Request for search *(Patents Form 9A)*:

Request for substantive examination
(Patents Form 10):

Any other documents:(please specify)

---

11   I/We request the grant of a patent on the basis of this application.

Signature(s):   Date:

---

12   Name, e-mail address, telephone,
fax and/or mobile number, if any,
of a contact point for the applicant:

Intellectual Property Office

Form TM3
Application to register a trade mark

Fee £200 [Includes one class of goods or services]
£50 [for each additional class]

Use this form to file your application to register a trade mark.

Do not use this form if you wish to pay a reduced fee, you can **save £30** by filing your application online.

††NOTE: *The details indicated with †† are displayed on the IPO website shortly after receipt. If the application is accepted, details are published on the internet in the Trade Marks Journal which is fully searchable by the public.*

1. **†† Full name**
   Proposed owner

   Owner type
   Specify whether Person, Registered Company/LLP, Partnership, Trust or Other

   †† Address
   If the address is not within the United Kingdom, European Economic Area (EEA) or the Channel Islands you must also complete section 2 below

   Entitlement to hold property
   Your trade may be declared null and void if you do not have the necessary standing to hold intellectual property. It is your responsibility to ensure that the applicant is legally capable of owning property in the name given on this application.

   Postcode

   Email address
   Complete if you have no representative and would like us to correspond with you by email.

   Company registration number
   Complete if the applicant is a company or LLP incorporated in the UK

   Country of incorporation
   If registered in USA also enter the 'State' e.g. 'Delaware'.

2. **†† Representative name**
   If you have no representative, go to section 3.

   †† Address
   The address provided in this section must be in United Kingdom, European Economic Area (EEA) or the Channel Islands.

   NOTE: *We will communicate with the representative, if this section has been completed.*

   Postcode

© Crown Copyright 2013   TM3

Email address
Complete if you would like us to correspond with you by email.

Representative type
Please tick appropriate box

IP Professional
In-House IP department
Lawyer/Solicitor
Other

3. Trade mark type
99% of applications are Trade Marks.
Certification Marks – indicate that goods or services meet a defined quality standard.
Collective – indicate the goods or services of a member of a trade association.

Trade Mark
Certification Mark
Collective Mark

4. Number of trade marks in series
Enter number only if applying for a series of trade marks (Max. 6 trade marks).

NOTE: 'Series' is a number of marks with very small differences e.g. 'danryvol', 'DANRYVOL', 'Danryvol'.

££ More than 2 marks are subject to an additional fee of £50 per

5. Representation of your trade mark or trade marks
Enter your trade mark in the space provided or attach on a separate sheet. Tick if attached

If your trade mark is a 3D shape and you are showing different views of the same mark please indicate the number of views in the box provided. The max number of images per trade mark is 6.

TM3

6. Trade mark classification

You need to tell us which goods and services you are going to use your trade mark for. Goods and Services are classified in an internationally agreed list of classes.
For information on how we classify: visit www.ipo.gov.uk/types/tm/t-applying/t-class.htm

You must tell us which class your goods and services belong in e.g. Class 25, Clothes. You can search and classify goods and services using the classification search tool 'TMCLASS'. Visit http://oami.europa.eu/ec2/

££ Applications with more than one class of goods or services are subject to an additional fee of **£50** for each additional class – use continuation sheet if necessary

Class Number	List of goods and services

7. Trade mark description (Optional)
If your trade mark is not a traditional trade mark such as a word, logo, picture, letters etc. You can tell us here. e.g. if it is a 3 dimensional shape or hologram.

8. Limitation (Optional)
Enter any limitations to your rights that you wish to volunteer. e.g. If you want to limit your rights to the trade mark to particular geographical areas of the UK.

9. Disclaimer (Optional)
If you want to volunteer to disclaim any rights to a part of your mark, you can do so here.

10. Priority claim (Optional)
If you have applied for this trade mark outside the UK in the last six months, you can claim priority by entering the details here.

Priority Claim Country

Application / Registration Number

Priority Claim Date

Priority claim type
You must tick only one of the priority claim type options

For All Goods and Services ☐

For Some of the Goods and Services ☐

11. Declaration
Warning! You cannot make any changes to the trade mark(s) applied for or add more goods or services to your application once it has been submitted. The application fee is non-refundable (even if the application is not acceptable).

I confirm that:

The applicant is entitled to hold property.

The terms used to describe the goods and/or services listed in this application should be given their ordinary and natural meaning.

The trade mark is being used by the applicant, or with his or her consent, in relation to the goods or services shown, or there is a bona fide intention that it will be used in this way.

I understand and accept that I cannot make any changes to the trade mark(s) applied for or add more goods or services once the application is submitted. I also accept that any application fees paid are non-refundable (even if the application is not acceptable).

Signature

Name
(BLOCK CAPITALS)

Date

Number of sheets attached to this form

12. Your reference
Complete if you would like us to quote this in communications with you, otherwise leave blank.

Contact details
Name, daytime telephone number of the person to contact in case of query.

Checklist

Please make sure you have remembered to:

☐ Sign and date the form

☐ Complete fee sheet (Form FS2)

☐ Enclose the fee and fee sheet. Make cheques payable to Intellectual Property Office

Where to send

Intellectual Property Office
Trade Marks Registry
Concept House
Cardiff Road
Newport
South Wales
NP10 8QQ

Schedule

Intellectual
Property
Office

Designs Form DF2A – Guidance Notes

Section 1 – Applicant Details

If you are applying as an individual you should enter your full name. Alternatively if the owner of the design is a company you should enter the 'full registered company name' here. 'Full Registered Company name' is the name of the incorporated Company or Limited Liability Partnership which will own the Design. In the UK, incorporated bodies usually have a formal designation such as 'Limited', 'Plc' or 'LLP'.

Partnership - You must enter the details of at least one partner. When completing the first partner's name please also type in the name of the partnership e.g. John Smith a partner in The Demo Partnership.

Trading name - A trading name is NOT the same as a company name. If using a trading name as part of this application; you must also provide the name of the body which is legally entitled to hold property e.g. 'Smith Ltd trading as Smith's'.

Please be aware that your design may be declared null and void if you do not have the necessary standing to hold intellectual property. It is your responsibility to ensure that the applicant is legally capable of owning property in the name entered on this application.

Section 2 – Representative Details

Please only complete this section if you have someone acting on your behalf such as an IP Attorney, solicitor etc. All correspondence will be sent to this address.

Section 3 – Number of Designs

Please specify the total number of designs applied for.

Section 4 – Fee Paid

Application to be published immediately: **£60** (for a single design or for the 1st design in a multiple).
£40 for each additional design if filing a multiple.

Publication to be deferred: **£40** (for a single design or for the 1st design in a multiple).
£20 for each additional design if filing a multiple.

Please use the designs fee calculator on the website to ensure the correct fees are paid as we cannot refund your application fee.

Section 5 – Do you wish to register your Design straight away?

You can choose to defer the registration of your application for up to 12 months from the date of filing your application. This allows you to secure a filing date whilst also giving you more time to market your product or to apply for a patent before publicly disclosing your design(s) on our website. The initial fees for a deferred application are less (see section 4 notes for fees). Please note that if you defer registration of the design, it will not be registered until you instruct us to do so. A further form and fee will be due within 12 months of your filing date. Failure to file the form DF2C and fee within the 12 month non-extendable deadline will result in your application being refused.

Section 6 – What is your Design?

You should state exactly what the design actually is e.g. logo, chair, car, t-shirt, wardrobe etc.

Section 7 – Illustrations of your Design

Illustrations should be presented as follows:
- Either line drawings, CAD or photographs showing the design as it actually appears to the eye
- Show the design in isolation e.g. with no other designs or background detail visible
- Plain A4 paper
- Do not include any technical detail such as borders/labelling/dimensions
- Show all relevant views of the design (maximum of 7) e.g. perspective view, rear view, front view, side view etc.
- The only text accepted on representations are view designations (perspective view, front view etc) and/or disclaimers should you choose to use them
- Please do not fold or crease the images
- Please ensure the print quality of the images is not effected with printer lines etc

Examples of acceptable illustrations are over leaf.

Section 8 – Do I need to include a disclaimer or limitation?

A disclaimer or limitation, alongside the illustration, can help define the scope of protection given to your design. It is important that you think carefully about whether you want or need to make a disclaimer. If your design contains features e.g. pattern(s) or colour(s) which do not relate to your design, then you should consider the use of a disclaimer. Equally, if your design only relates to a part of the product, then you should consider whether or not to limit the scope of your design to that particular element.

Disclaimers and limitations can play an important role in any dispute where you want to enforce your rights against someone who is copying your design, or if there is any dispute about its validity.

Here are three example scenarios where a disclaimer or limitation could be used:

Disclaimers

Example 1: You want to file a design for a bike. Although the illustration shows the bike in pink, the design relates only to its *shape*, and not to the colour shown in the illustration. In this case, the applicant should consider making a disclaimer to the colour shown in the illustration. If they do not, the protection given may be limited to the design shown in the colour pink. In this example, the designer might say *"no claim is made to the colour shown in the illustration"*.

Example 2: You want to file a design for a teapot. Although the illustration shows a pattern applied to the teapot, the design relates only to its shape and not to the pattern. In this case, a feature of the design (the pattern) is *not* part of the design and so the designer should consider disclaiming the pattern shown in the illustration. In his example, a disclaimer might be *"No claim is made for the pattern shown on the surface of the teapot"*.

Limitation

Example 3: You want to file a design for a table. Although the illustration shows the whole table including its top, bottom, legs and side views, the design relates to the legs of the table only - the table top is *not* part of the design. The designer should consider limiting the design to the table legs. In this example, you may wish to highlight the relevant part of your illustration (the table legs) by using a clear visual border, or by greying-out the irrelevant parts of your design.

If I need to record one, how do I submit a disclaimer or limitation?

You can specify the part of your design which you want protection for by using either a *disclaimer* or a *limitation* as outlined in the above examples.

A disclaimer and limitation can either be shown visually (graphically) by indicating that protection in intended for only part of your article. Using the table example, you may wish to highlight the relevant part of your illustration (the table legs) by using a clear visual border, or by greying-out the irrelevant parts of your design.

Alternatively, If you wish to record a written disclaimer or limitation (and it isn't already clear from the illustrations you have submitted), you should enter it in **Section 8** of the form:

Please note that disclaimers and limitations are optional. If you want protection for the whole of your design as shown on the illustration(s), then you should not use them. Please also note that disclaimers and limitations will limit the scope and extent of protection conferred upon your design."

Section 9 – Repeated Surface Pattern (optional)

Views of designs with a repeated surface pattern should show the complete pattern and be surrounded with enough of the repeat to fully illustrate the entire pattern. Examples of such designs are wallpaper or textile materials.

Additional Information

On filing your application, it will undergo an examination process where we will examine your application to ensure it meets the requirements of the Registered Designs Act 1949 and The Registered Designs Rules 2006. Please be aware that we do not conduct a novelty search as part of the process. We can conduct a search of identical designs registered in the UK on request by filing form DF21 and a £25 fee.

The rules regarding ownership have changed (1st October 2014). Employers continue to own the rights to any design created by employees as part of their employment. If you are an independent designer you own the rights to your design even if commissioned unless a contract is in place stating otherwise. If you pay someone commission to create a design on your behalf, you do not own the rights in the design unless a contract is in place stating otherwise.

Section 7 cont. - Illustrations dos and don'ts.

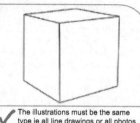

✓ Show one view per upload (ie a file must not contain more than one view of the design)

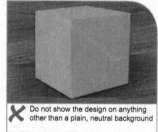

✓ The illustrations must be the same type ie all line drawings or all photos and not a combination

Designs with a repeated surface pattern should include the complete pattern and enough of the repeat to show how the pattern repeats

✗ A file must not contain more than one view of the design

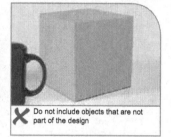

✗ The illustrations must not contain a combination of line drawings, prints or photos

✗ Do not upload a repeated pattern on its own (ie you must show it being repeated)

✗ Do not upload any flags, crowns, royal emblems, anything illegal or immoral

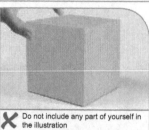

✗ Do not show the design on anything other than a plain, neutral background

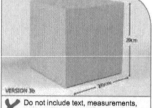

✗ Do not include objects that are not part of the design

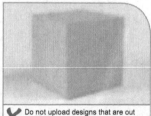

✗ Do not upload designs that are out of focus

✗ Do not include any part of yourself in the illustration

VERSION 3b

✗ Do not include text, measurements, technical features, etc

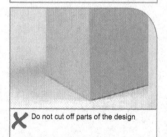

✗ Do not cut off parts of the design

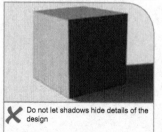

✗ Do not let shadows hide details of the design

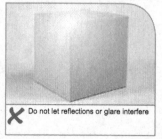

✗ Do not let reflections or glare interfere

Intellectual
Property
Office

Designs Form DF2A
Use this form to file your application to register a Design

Fee - Up to date fees are published on our website (www.gov.uk/government/publications/design-forms-and-fees)

As part of the registration process your name, address and details of your design are published.

1. **Full name**
   Proposed owner (as it will appear on the Registration Certificate)

   Owner type
   Note: Specify the owner is an individual, registered company/LLP, partnership or trust

   Address
   If the address is not within the United Kingdom, European Economic Area (EEA) or the Channel Islands you must also complete section 2 below

   Postcode

   Email address
   Complete if you have no representative and would like us to correspond with you by email.

   Country of incorporation
   If registered in USA also enter the 'State' e.g. 'Delaware'.

   Contact details
   Name, daytime telephone number of the person to contact in case of a query

2. **Representative name**
   If you have no representative, go to section 3.

   Address
   The address provided in this section must be in the UK, European Economic Area (EEA) or the Channel Islands.

   Postcode

   Email address
   Complete if you would like to correspond via this address

3. **Number of Designs**
   Please confirm the number of designs you are applying for.

4. **Fee Paid**
   Please ensure the correct fee has been enclosed.
   A fee calculator is available with this form
   (https://www.gov.uk/government/publications/design-application-fee-calculator)

About Your Design

This is the ____ (for example, first) design out of a total of ____ designs.

You must answer these questions for each design in a multiple application, so copy this sheet as many times as you need.

5. Do you wish to register your Design straight away?
Please indicate **yes** or **no** here

Yes ☐   No ☐

If you wish to delay registration, detailed guidance notes as to why this may be a possible option for you are available with this form

6. What is your Design?
Please tell us what the design is i.e. a chair, wardrobe, hosepipe etc

7. Representation of your Design
Please include a representation of your design on the representation sheet provided.
Further details on how to present your representations are available with this form.

8. Disclaimers and Limitations (Optional)
If you want to disclaim any part of the design e.g no claim is made for the colours shown, you can do so here. This should also be applied to the representation of your Design

9. Repeated Surface Pattern (Optional)
Write "RSP" if this is the design of a pattern which repeats across the surface of a product, for example, wallpaper here and on your representation of the Design

10. Priority Claim (Optional)
If you have applied for this Design outside of the UK in the last six months you can claim priority by entering the details here

Priority Claim Country	
Application/Registration Number	
Priority Claim	

11. Declaration

I can confirm that:

The applicant is entitled to hold property

I understand and accept that any application fees paid are non-refundable (even if the application is not acceptable)

Signature
A signature must be supplied

Name
(BLOCK CAPITALS)

Date

12. Your reference
Complete if you would like us to quote this in communications with you, otherwise leave blank.

Checklist

Please make sure you have remembered to:

☐ Sign and date the form

☐ Complete <u>fee sheet</u> (Form FS2)

☐ Enclose the fee and fee sheet

☐ Enclose the representations of the Design

Where to send

Intellectual Property Office
Trade Marks Registry
Concept House
Cardiff Road
Newport
South Wales
NP10 8QQ